Stereology for Biological Research
With a Focus on Neuroscience

Jack Glaser, Geoff Greene, Susan Hendricks

© Copyright MBF Press 2007. All rights reserved.

Published by MBF Press, 185 Allen Brook Lane, Suite 201, Williston, VT 05495, USA.

Printed in the United States of America

Neurolucida, Stereo Investigator, Virtual Slice, and Virtual Tissue are trademarks or registered trademarks of MicroBrightField, Inc.

Other company and product names mentioned herein are the trademarks or registered trademarks of their respective owners.

No part of this book may be reproduced, in any form or by any means, without permission in writing from the publisher.

ISBN: 0-9786471-0-6

Library of Congress Preassigned Card Number: 2006930035

First Printing

Table of Contents

Acknowledgements v
Preface vii
PART I How Design-Based Stereology Can Help Advance Your Research 1
PART II Prerequisites for Design-Based Stereological Studies................................ 13
    Can I initiate design-based stereological investigations on my existing tissue sections? 15
    Is it correct that the application of design-based stereology requires the use of thick tissue sections? How thick must the sections be? 15
    My tissue sections require a certain plane of section to identify the region of interest. Does this impair the applicability of design-based stereological methods? 16
    Can I perform design-based stereological analyses on my existing standard laboratory microscopes, or is special equipment necessary?............................... 16
    What does computer-based stereology offer? 17
    How expensive is such a semiautomated, computer-based stereology system?.. 18
PART III Design-Based Stereological Methods 19
    How can I quantify the volume of my region of interest?.................................... 21
    How can I quantify the total number of cells in my region of interest?............. 23
    How can I measure the size of the cells in my region of interest? 33
    How can I determine the length density and total length of capillaries or cell processes in my region of interest? 36
    How can I determine the spatial distribution of cells in my region of interest? ... 40
    What else can I do? 44
    How can I assess the variation of estimates obtained with design-based stereology? 45
    How does consideration about variability affect the implementation of design-based stereological estimators? 50
PART IV Application of Design-Based Stereology in Biological Research................ 53
    Application of design-based stereology in research on brain aging and Alzheimer's disease 55
    Application of design-based stereology in schizophrenia research..................... 57
    Application of design-based stereology in stem cell research 58
    Application of design-based stereology in research on the lung, kidney and placenta 59
PART V The Future of Design-Based Stereology... 63
    Development of novel design-based stereological methods 65

    More precise prediction of the variability of stereological estimates 67
    Integration of design-based stereology with anatomical mapping 67
    Integration of design-based stereology with confocal microscopy 68
    Integration of design-based stereology with digital representations of histological tissue specimens 72

PART VI Further Reading 77
    Books dedicated to design-based stereology: 79
    Reviews dedicated to design-based stereology: 80

Part VII Glossary 83

INDEX 91

About the Authors 93

Acknowledgements

The authors gratefully acknowledge the guidance and knowledge generously provided by Dr. Christoph Schmitz (Department of Psychiatry and Neuropsychology, Division of Cellular Neuroscience, Maastricht University, Maastricht, The Netherlands) All histological photographs were taken from ongoing research in his laboratory. The corresponding animal experiments were approved by the relevant Institutional Review Boards. The postmortem human brain was provided to Dr. Schmitz by Dr. Helmut Heinsen (Morphological Brain Research Unit, University of Wuerzburg, Wuerzburg, Germany). The corresponding autopsy was performed more than 10 years ago after consent was obtained from a relative according to the laws of the Federal Republic of Germany. The use of this autopsy case for scientific investigations as outlined in this publication has been approved by the relevant Institutional Review Boards.

The illustration on page 49 is reprinted from Neuroscience, 136, L. Slomianka and M.J. West, "Estimators of the Precision of Stereological Estimates: An Example Based on the CA1 Pyramidal Cell Layer of Rats", page 757, Copyright 2005 with permission from Elsevier.

Preface

In recent years design-based stereology has taken its place as the standard methodology for quantitative histology used in biological research, particularly in the identification of subtle—yet important—alterations in the morphology of the central nervous system based on disturbances during development or in neuropsychiatric and neurodegenerative diseases. This importance is reflected by many excellent books and reviews dedicated to design-based stereology and its applications.

This book is not just another addition to the list. Rather it aims to take you on a brief journey to discover the world of design-based stereology seen through the eyes of a scientist not familiar with, but interested in, the application of this methodology. During your journey, you will visit the following places of interest:

- In **Part I** you visit the lab of a neuroscientist who has just discovered the advantages of design-based stereology over less sophisticated approaches in quantitative histology.

- In **Part II** you learn from experts in the field about the prerequisites needed for design-based stereological studies.

- **Part III** provides you with a short overview of the principles of design-based stereological methods.

- In **Part IV** you will hear from scientists who actually have implemented design-based stereology in their laboratories, and will find out how this has helped to advance neuroscience and other fields of biological research.

- In **Part V**, perhaps the highlight of the trip, you will have a chance to observe the work of laboratories engaged in implementing the future of design-based stereology.

- In **Part VI**, at the end of the journey, you will find information about further reading.

PART I

How Design-Based Stereology Can Help Advance Your Research

Stereology is a set of methods designed to ensure rigorous quantitative analysis of the size, shape, and number of objects. When properly used, stereology plays an important role in validating and rejecting experimental hypotheses in biological research. It produces results that are accurate, efficient, and more reliable than other *ad hoc* quantitative analyses. Stereology provides an important contribution to the advancement in biological research by improving the consistency and dependability of quantitative analytical results produced in the laboratory and reported in scientific publications.

Application of stereological methods to biological studies permits researchers to effectively and efficiently gather data. To better understand the principles of design-based stereology and its advantages over less sophisticated approaches in quantitative histology, let us start with an example.

Example: counting Purkinje cells in the rat cerebellum

Let us assume that a researcher wanted to test the hypothesis that a certain treatment results in loss of Purkinje cells in the cerebellum of rats. To test this hypothesis, the researcher generated 5 μm thick paraffin sections through the cerebellum of six treated and six untreated rats and stained them with cresyl violet (a Nissl stain). Then she analyzed one or two representative sections per animal under a standard microscope with a 40x lens and a grid in one eyepiece defining a certain area. Choosing 10 microscopic fields per animal that showed the best staining of the cerebellar Purkinje cell layer, she then counted all Purkinje cells she could identify within the area defined by the grid in the eyepiece. Afterwards she divided the total number of Purkinje cells counted per animal by the total area examined (the size of the area defined by the eyepiece grid multiplied by 10, the number of fields examined per animal). From these data, she calculated the mean Purkinje cell density in the treated and the untreated animals and compared the groups with Student's t test. The analysis showed a slight but statistically non-significant ($p > 0.05$) reduction in the mean Purkinje cell density in the treated rats compared to the untreated animals. The researcher concluded that the treatment did not result in Purkinje cell loss in the rat cerebellum.

Numerous studies of this style can be found in the literature, raising the question of why anyone would suspect that something could be wrong with this analysis, and if not, why there would be a need to perform such a study differently. Indeed, there is nothing procedurally wrong with the counting in this example, even from a rigorous stereological point of view.

Instead, the discussion should focus on whether the data analysis procedure will accurately demonstrate a difference between groups if it exists. Or, expressed another way, the methodology should be approached such that the failure to reject the Null hypothesis is the accurate biological conclusion to be reached (i.e., that the treatment did not result in loss of cerebellar Purkinje cells). Large uniform differences between groups are likely to be observed with this counting method, but subtle changes in number or density are more likely to be missed. To understand the potential problems with this method, one has to consider that the study was performed on histological sections from brain tissue, and might therefore be influenced by the following issues: bias in the choice of "representative" sections, reliability, and validity of cell counts and density measurements. We will examine each of these issues in turn and how they are resolved by design-based stereology:

- The classification of a certain section through a given region of interest as "representative" implies that the results obtained from this section are more or less identical to the average of the results that would have been obtained if all sections encompassing the region of interest were analyzed in the same way.

  This was not tested in the above example. Rather it has been shown in the literature that results from cell counts obtained on single "representative" sections (defined by anatomical landmarks or quantitative pre-analysis) through the cerebellum and hippocampus of rats can differ considerably from cell counts obtained on *systematically* and *randomly sampled* (SRS) series of sections encompassing these brain regions.[1] In general terms, the analysis of an SRS series of sections encompassing the entire region of interest has become the gold standard in design-based stereology.[2] The principle of how to generate an SRS series of sections through the rat cerebellum can be seen in Figure 1, part A.

---

[1] Schmitz et al., Neuroscience 2005;130:935.
[2] For an excellent introduction into the concept of systematic-random sampling in design-based stereology see Gundersen and Jensen, J Microsc 1987;147:229.

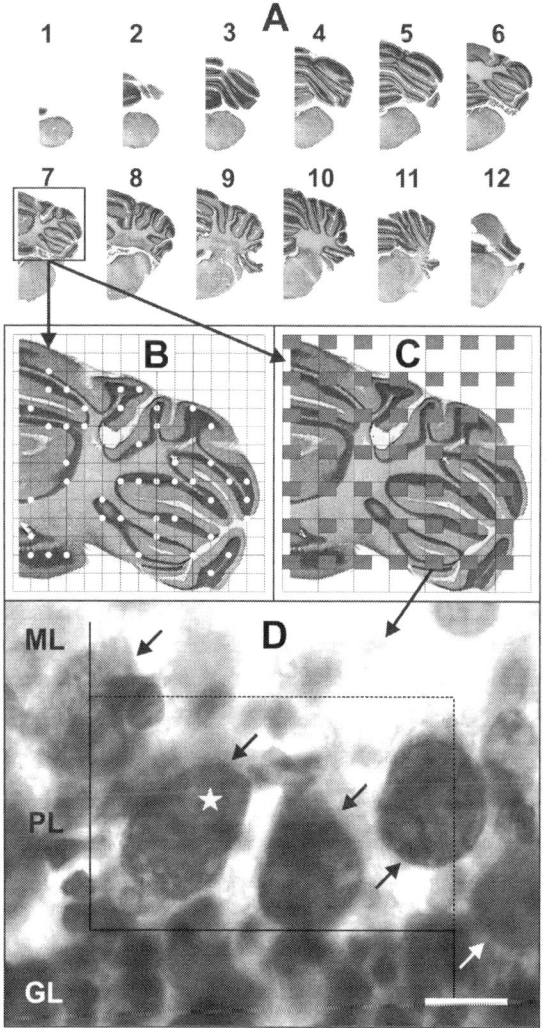

Figure 1. Essential steps in the design-based stereological analysis of a brain region of interest (here: the right cerebellar half of a 9-month-old rat. The rat was killed under deep anesthesia by transcardial perfusion with 4% paraformaldehyde in phosphate buffered saline (PBS, pH 7.4). Then the brain was removed from the skull and postfixed in the same fixative at 4° C for 2 weeks). **A,** Systematically-randomly sampled (SRS) series of 50 μm thick, Nissl-stained coronal cryostat sections encompassing the entire right cerebellar half, with a distance of 250 μm between the upper surfaces of the sections (i.e., every fifth section was selected [systematic aspect of sampling], and section no. 1 was randomly selected from the first 10 sections of the cerebellar half [random aspect of sampling]). **B,** Section no. 7 from the series, shown at higher magnification, superimposed with a rectangular lattice with uniform distances between the lines in directions X and Y. The position of the lattice on the section

was randomly selected. As a result, the intersections of the lattice and the cerebellar granule cell layer (marked with points) were systematically and randomly generated (systematically because of the uniform distances between the lines of the lattice in directions X and Y, and randomly because of the random position of the lattice on the section). From the total number of points (here: 45) and the distance between the lines in directions X and Y an estimate of the projection area of the cerebellar granule cell layer in this section is obtained. Summing up the corresponding estimates from all sections and multiplying this value with the uniform distance between the sections (here: 250 µm) gives an estimate of the total volume of the granule cell layer in the investigated cerebellar half (Cavalieri's principle). **C,** The same section as in B, randomly superimposed by another rectangular lattice with uniform distance between the lines in directions X and Y. This lattice defines the positions of a systematically and randomly selected set of microscopic fields (indicated as gray rectangles) with uniform distances between the fields in directions X and Y, at which the section is then inspected at higher magnification to perform 'local estimates' (i.e., cell counts as well as estimates of cell size, density of lengthy biological objects [such as capillaries or cell processes] and the spatial distribution of cells). **D,** High-power representation of the microscopic field indicated in C, showing details of the cerebellar molecular layer (ML), Purkinje cell layer (PL) and granule cell layer (GL). Five Purkinje cells are found in this microscopic field (arrows; the cell marked by the asterisk is also displayed in Figure 3 and Figure 4). To count the Purkinje cells in an unbiased manner (i.e., irrespective of their size, shape and spatial orientation) the microscopic field is superimposed with an *unbiased counting frame*, consisting of two *exclusion lines* (solid lines in this example) and two *inclusion lines* (dotted lines in this example). The mode of operation of the unbiased counting frame is explained in Figure 3 to Figure 5.

The pictures shown in A were each generated from up to 17 images captured with a 1.25x lens (Olympus Plan Apo; NA = 0.04), made into one montage using MBF's Virtual Slice module of the Stereo Investigator software. Scale bar represents 8 mm in A, 2.5 mm in B and C, and 20 µm in D.

- Also problematic is the selection of certain microscopic fields based on local staining quality (or any other subjective criteria). Classifying a certain set of microscopic fields per section as "representative" implies that the results obtained on these microscopic fields are more or less identical to the average of the results that would have been obtained if all regions of the section were analyzed in the same way.

  Again, this was not tested in the example. To avoid this, performing an analysis on systematically and randomly sampled sets of microscopic fields covering the entire region of interest will ensure that results obtained are representative of the entire structure. The principle of how to generate an SRS set of microscopic fields on a section from the rat cerebellum is shown in Figure 1, part C.

- In paraffin sections, cerebellar Purkinje cells show perikarya with a mean diameter of approximately 20 µm. In the above example, our researcher actually counted Purkinje cell fragments (also referred to as cell profiles) instead of Purkinje cells because the tissue was sectioned at 5 µm. However, note that the larger the size of a cell, the higher the probability that a fragment of this cell is found in more than one section from the corresponding tissue.[1]

  This might not have had any impact on the study outlined in the example, provided that the treatment did not influence the mean perikaryal size of the cerebellar Purkinje cells. But since it was not explicitly tested, we do not know for sure. In general terms, analyses of cell densities or absolute cell numbers within a given tissue volume based solely on counts of cell profiles in the sections might not represent the true density or number of cells within this tissue volume (and might therefore be biased). The solution to this problem in design-based stereology is to count a unique punctuate feature of cells (such as the cell 'top') within thick tissue sections with unbiased virtual counting spaces, in order to make cell counts that are independent of cell size as well as shape and spatial orientation (shown in Figure 1, part D, and addressed in detail in Figure 3 to Figure 5 below).

- Finally, the use of cell density measurements in studies that test the hypothesis of alterations in cell numbers can be problematic. Perhaps one of the most intriguing examples in this regard is the finding that subjecting pregnant rats to a certain kind of X-ray-irradiation caused a reduction in the mean total numbers of hippocampal pyramidal and

---

[1] For illustration see Figure 3 in Schmitz and Hof, Neuroscience 2005;130:813.

granule cells as well as of cerebellar Purkinje and granule cells in the brain of the offspring by approximately 50%, without any alteration in the mean densities of these cell types.[1] It is important to emphasize this finding again in another way—there was no observed difference in the cell density, yet there actually was a 50% reduction in cell number. Density measurements would not have found this reduction because it was due to a concomitant reduction in the mean size of the corresponding cell layers in the rat brain by approximately 50% caused by the prenatal X-ray-irradiation paradigm. Again, this phenomenon might not have had any impact on the study outlined in the example, provided that the treatment did not influence the size of the cerebellar Purkinje cell layer. This, however, was not tested in the example. Since design-based stereology provides an estimate of the total number of cells independent of volume, this confound can be avoided.

A key to unbiased estimation using stereology is the requirement that each object within the region has the opportunity to be counted once and only once. To ensure this, SRS relies on randomized start and placement of the sampling paradigm and regular sampling of the structure: 1) through the entire extent of the tissue by selection of sections to sample, 2) across the surface of the section, and 3) within the thickness of the section. Integration of proper sampling methodology in your experimental design is the key to success with design-based stereology.

Design-based stereology provides several different methods to estimate total numbers of cells (or total numbers of any other countable biological objects of interest). With the first one (called the Optical Fractionator[2] as described on page 24), you estimate the total numbers of cells directly from the number of cells counted with an SRS set of unbiased virtual counting spaces covering the entire region of interest in a three-dimensional manner. The second method, called $N_V \times V_{Ref}$ method, combines an estimate of the mean cell density in this region (N_V) with an estimate of the volume of the reference space of a region of interest (V_{ref}).[3] The Optical Fractionator technique is generally easier and faster. Both techniques use a uniform distance between the unbiased counting spaces in directions X, Y and Z.

All aspects addressed so far deal with the validity of results from quantitative histological studies. But what about the reliability of results from

---

[1] Schmitz et al., Neuroscience 2005;130:935. Other examples in the literature showing discrepancies between alterations in cell densities and alterations in corresponding total numbers of cells are also reviewed in this study.
[2] Described in the literature for the first time in West et al., Anat Rec 1991;231:482.
[3] Described in the literature for the first time in West and Gundersen, J Comp Neurol 1990, 296:1.

such studies? What could have been the outcome in the example if the researcher would have selected other sections for analysis, different microscopic fields to count cells, or both? Actually, one has to admit that this question simply cannot be answered; the results might well have been more or less the same.

One of the most powerful developments in design-based stereology has been the introduction of statistical methods for predicting the precision of the calculated estimate of the total numbers of cells and of the estimated volumes of regions of interest.[1] Computer simulations have demonstrated the validity and reliability of these prediction methods.[2] This has given researchers the possibility of creating "what if?" scenarios if, for instance, they would have selected another SRS series of sections for analysis, another SRS set of microscopic fields to count cells with unbiased virtual counting spaces, or both. As opposed to profile counting, design-based stereology provides a means to determine how well your estimate models the true population.

In summary, design-based stereology provides a powerful set of methods, focusing on just one single aim: improving the validity of researchers' biological conclusions from quantitative histological investigations. Applying design-based stereology in the study outlined in the example would have resulted in the following analysis:

A better way to count Purkinje cells in the rat cerebellum

Let us assume, once again, that our researcher wanted to test the hypothesis that a certain treatment results in loss of Purkinje cells in the cerebellum of rats. To test this hypothesis, she generated an SRS series of every fifth, 50 μm thick coronal cryostat section encompassing the entire cerebellum of six treated and six untreated rats and stained them with cresyl violet such that the final mounted section thickness was 40 μm. Then she analyzed this SRS series of sections with a stereology workstation, comprised of (among other components) a microscope, a color digital CCD camera, a three-axis computer-controlled stepping motorized specimen stage and stereology software (Figure 2). After tracing the boundaries of the cerebellum on each section at low magnification (i.e., with a 1.25x lens) on video images displayed on the monitor, the stereology software generated an SRS sampling set covering the entire cerebellum on each section. The software

---

[1] For details see, e.g., Gundersen and Jensen, J Microsc 1987;147:229; Glaser and Wilson, J Microsc 1998;192:163; Schmitz, Anat Embryol 1998;198:371; Gundersen et al., J Microsc 1999;193:199; Schmitz and Hof, J Chem Neuroanat 2000;20:93; Cruz-Orive and García-Fiñana, J Microsc 2005;218:6.
[2] Ibid.

drove the motorized stage to each sampling location for inspection and counting at a high magnification (40x). As the researcher focused through the tissue, the software provided warnings whenever she intended to count a Purkinje cell that did not fulfill the criteria for unbiased counting in three dimensions. After the analysis was completed for all sections from a given animal, the software automatically calculated the estimated total number of Purkinje cells for this animal, together with a prediction of the precision of this estimate. From this data, the researcher calculated the mean total numbers of Purkinje cells in the treated and the untreated animals and compared the groups with Student's t test. The analysis showed for the treated rats a slight (20%) but statistically significant ($p < 0.05$) reduction in the mean total number of cerebellar Purkinje cells compared to the untreated animals. The researcher concluded that the treatment resulted in Purkinje cell loss in the cerebellum of rats. In the resulting publication, she could demonstrate that the results of this study were obtained in a manner independent from the size, shape, and spatial orientation of the investigated cells, as well as from the sections and microscopic fields that were analyzed. In addition, the researcher could demonstrate that the observed difference of approximately 20% in the mean total numbers of cerebellar Purkinje cells between treated and untreated animals actually represented a biological difference between the groups rather than a difference mainly caused by sampling variability during the quantitative histological investigation.

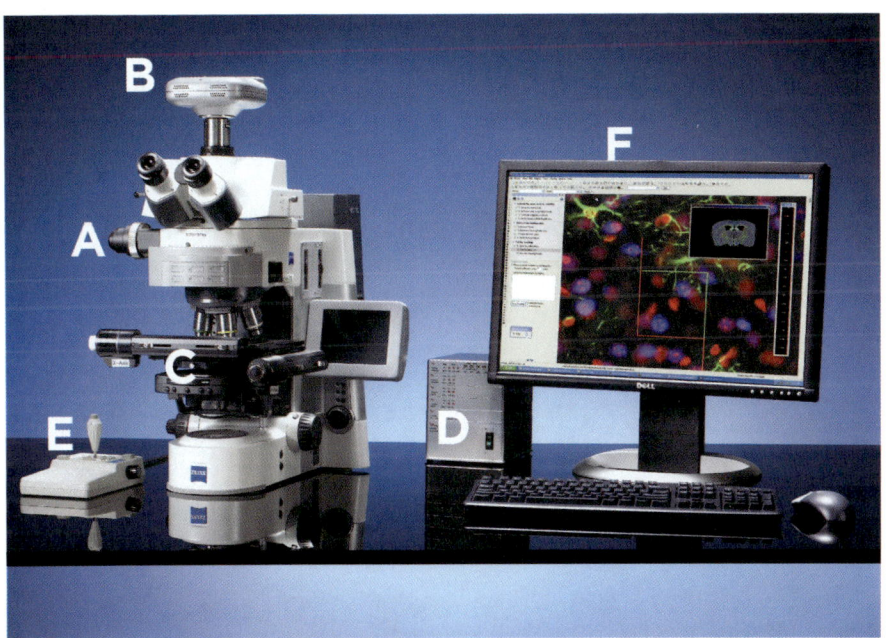

Figure 2. A typical MBF Bioscience Stereo Investigator system. **A**, Research microscope. **B**, Color digital CCD camera. **C**, Two-axis computer-controlled stepping motor specimen stage. **D**, Control unit. **E**, Three-axis joystick. **F**, PC workstation with video card, stereology software and LCD monitor. The linear Z-axis position encoder is not shown.

PART II

Prerequisites for Design-Based Stereological Studies

Researchers often ask the following questions when they begin to consider design-based stereology as a useful methodology in their research:

Can I initiate design-based stereological investigations on my existing tissue sections?

To address this question, let's look at the example given in Part I. If you are interested in the volume of a particular region of interest and/or the total number of cells in this region, and you want to get valid estimates and information about the precision of these estimates, you must have a systematically and randomly selected (SRS) series of sections encompassing the entire region of interest (as shown in Figure 1, part A). Analyses of single, "representative" sections through the region of interest will lead to potentially biased results with unknown precision.[1]

Is it correct that the application of design-based stereology requires the use of thick tissue sections? How thick must the sections be?

For volume estimates with the Cavalieri estimator, the sections can be thin (such as regular 5 µm thick paraffin sections). To analyze total numbers of cells using the Optical Fractionator, however, the sections should have a certain thickness that allows the placement of unbiased virtual counting spaces within the section thickness (details are provided in Figure 3 and Figure 4). The same applies for analyses of the length of tubular objects such as capillaries or cell processes (using protocols such as Space Balls [Figure 7] or Isotropic Virtual Planes)[2] or the spatial distribution of cells in a given region of interest using the Nearest-Neighbor Distance Distribution Function (Figure 8 and Figure 9).[3] Furthermore, the sections must be free from non-uniform compression in the Z axis[4] and should have a smooth surface.[1]

---

[1] For a recent example see Schmitz et al., Neuroscience 2005;130:935. From a theoretical point of view, the sections to be analyzed for the purpose of determining the volume of a region of interest and/or the total number of cells in a region **must** fulfill the following criteria: First, you need access to the entire region of interest; otherwise you can make statements only for the parts of the region you actually have access to. Second, the entire region (or the cells in this region) must be recognizable with an appropriate staining method; otherwise you must limit the analysis to the parts which can be recognized. Third, all parts of the region of interest (or all cells in this region) must have the same chance to contribute to the sample selected for design-based stereological analysis; otherwise your statements will only be valid if you know how well each part of the region (or the cells in this region) are represented in the estimates.

[2] For a recent example, see Kreczmanski et al., Acta Neuropathologica 2005;109:510.

[3] Schmitz et al., Cereb Cortex 2002;12:954.

[4] Von Bartheld, Histol Histopathol 2002;17:639.

For investigations on the mouse and rat brain, 30 to 50 μm thick cryostat sections fulfill all these requirements and therefore appear optimally suitable for design-based stereological analyses.[2] For investigations on human or monkey brain, first determine whether blocks of tissue should be cut from the brains and histologically processed, or if sections from the entire hemisphere (or even the entire brain) shall be prepared. The cut section thickness can then vary between twenty and several hundreds of microns, depending on the scientific question and the individual staining protocol applied.[3]

My tissue sections require a certain plane of section to identify the region of interest. Does this impair the applicability of design-based stereological methods?

The appearance of biological tissue under the microscope always depends on the plane of section (for instance, consider the difference between cross-sections and longitudinal sections through the spinal cord). Accordingly, most biological tissue is not isotropic. This anisotropy (non-isotropic, i.e., having a preferential direction), however, can influence the results of quantitative histological analyses (for instance, in a cross-section through the spinal cord more nerve fiber transections [i.e., cross-sectional fiber profiles] are found than in a longitudinal section). In the earlier days of design-based stereology, this problem was addressed by introducing isotropy in the planes of section, i.e., by preparing "isotropic uniform random" sections[4] or "vertical" sections.[5] These methods, however, could rarely be successfully introduced in the field of neuroscience. Rather the issue of anisotropy in biological tissue was re-addressed by introducing isotropy in a novel set of design-based stereological methods that are outlined below. Now design-based stereological methods can be successfully applied on tissue sections with any plane of section.

Can I perform design-based stereological analyses on my existing standard laboratory microscopes, or is special equipment necessary?

In general, you can perform analyses on a volume of a region of interest with the Cavalieri estimator and point counting (details are provided in Figure 1), as

---

[1] Dorph-Petersen et al., J Microsc 2002;204:232.
[2] Note that certain investigations such as the determination of the spatial distribution of biological objects in a given region of interest with the Nearest-Neighbor Distance Distribution Function method might require even thicker tissue sections; see also Schmitz et al., Cereb Cortex 2002;12:954.
[3] See, for example, Perl et al., J Chem Neuroanat 2000;20:7; Heinsen et al., J Chem Neuroanat 2000;20:49;Schumann and Amaral, J Comp Neurol 2005;491:320.
[4] Miles and Davy, J Microsc 1976;107:211.
[5] Baddeley et al., J Microsc 1986; 142:259.

well as estimates of the total number of cells in a given region using the Physical Fractionator with two additions to most existing microscope setups. You will need a grid in one eyepiece and a specimen stage equipped with a precise vernier scale. With the addition of a Z position encoder it may also be possible to implement the Optical Fractionator to estimate cell populations (details are provided in Figure 1 and Figure 4). In contrast, analyses of the size of cells with the Nucleator, the Rotator, or Point Sampled Intercepts (details are provided in Figure 6) require a camera attached to the microscope at a minimum. Furthermore, investigations of the length of tubular objects such as capillaries or cell processes using the Space Balls method (details are provided in Figure 7) or the spatial distribution of cells in a given region of interest using the Nearest-Neighbor Distance Distribution Function method (details are provided in Figure 8) require the use of semiautomated, computer-based stereology systems as shown in Figure 2.

What does computer-based stereology offer?

Given the difficulties involved with manual stereology, it was the advent of semiautomatic, computer-based stereology systems that allowed design-based stereology to become a practical laboratory method. Specifically, the advantages provided by design-based stereology are best obtained when it is integrated into advanced, computer-based microscopy systems that optimize the collection, storage, and analysis of data. Such systems reduce both the observer's effort and potential errors associated with the use of the different design-based stereological methods. In addition, these systems offer the combination of computer-based anatomical mapping and rigorous stereological estimates.

Modern semiautomatic, computer-based stereology systems integrate a 3-axis, computer-controlled motorized specimen stage with a linear Z-axis position encoder, a digital camera, and a computer in order to acquire data from three-dimensional (3D) structures and implement the stereological methods mentioned above (as shown in Figure 2). The motorized stage is used to map brain regions and objects that are larger than a single microscopic field-of-view, to rapidly access specific locations throughout the entire region of interest regardless of the optical magnification, and to perform systematic random sampling.[1] The linear Z-axis position encoder accurately measures the actual focal position of the microscope stage. The tissue specimens are usually viewed on a computer monitor via a high-resolution color digital camera that captures at least 10 frames per second, allowing real-time visualization while focusing through the tissue. Furthermore, state-of-the-art stereology systems provide image file readers that are capable of accepting 3D confocal and MRI

---

[1] Glaser and Glaser, J Chem Neuroanat 2000,20:115.

image sets, as well as file formats generated by a variety of electron microscopes and flatbed scanners.

To perform a design-based stereological analysis, the user views the tissue specimen at each sampling site with the stereological method's geometric probe overlay superimposed upon it. For instance, when counting cells with the Optical Fractionator, the system displays an unbiased counting frame superimposed on the specimen (as shown in Figure 1, part D and Figure 3 to Figure 5). Then the user clicks on the sites of interaction of the specimen with the method's probe. The system saves all of the acquired data from all sampling sites, for all sections, for a given animal, in a file containing the 3D locations of all points entered into the file and their relationship to one another, i.e., as a contour or a set of points of a particular type. Once the tissue specimen has been analyzed, the software automatically calculates the final result (e.g., the estimated total number of cells in the region of interest), together with a detailed prediction of the precision of this estimate. These results and analyses can be prepared for publication, shared with other researchers, and saved for use with other software programs.

How expensive is such a semiautomated, computer-based stereology system?

There is no single answer to this question. The MBF Bioscience approach is to design and set up computer-based stereology systems custom-tailored to the user's exact scientific needs, integrating existing hardware such as microscopes, motorized stages, cameras, computers, etc. as much as technically possible.

A stereology system such as Stereo Investigator can be added to existing microscope research hardware or supplied as a turnkey system. For the stereology software alone, prices range from approximately $5,000 to $15,000 USD. For a system including a precision motorized stage, a high-resolution digital camera, computer, and software, prices range from approximately $25,000 to $40,000 USD, not including a research grade microscope. At the high end, complete turnkey multi-channel confocal stereology systems using structured illumination, such as a spinning disk, range from $175,000 to $350,000 USD. These prices are based on the 2006 MBF Bioscience price list. Costs may vary in different countries depending on importation duties and dealer pricing. MBF Bioscience can provide price-customized quotations with detailed line items for systems to meet a researcher's specific needs.

PART III

Design-Based Stereological Methods

This part of the book presents a short overview of the most relevant design-based stereological methods for biological research. The specific methods are illustrated by the following examples from the neurosciences:

- the volume of the granule cell layer in the right cerebellar half of a rat brain

- the total number of Purkinje cells in the right cerebellar half of a rat brain

- the mean perikaryal size of Purkinje cells in the right cerebellar half of a rat brain

- the capillary length density and total capillary length in the lateral nucleus of the amygdala in the right hemisphere from a human postmortem brain

- the spatial distribution of neurons in the lateral nucleus of the amygdala in the right hemisphere from a human postmortem brain

How can I quantify the volume of my region of interest?

The method of choice to quantify the volume of a region of interest is the Cavalieri Estimator protocol (based in large part on Cavalieri's principle[1]). Before performing a Cavalieri estimate, you must have access to a systematically and randomly sampled (SRS) series of sections encompassing the entire region of interest, with a uniform distance between the sections.

> The general concept of SRS series of sections encompassing a region of interest is shown in Figure 1, part A, depicting an SRS series of 50 µm thick Nissl-stained coronal cryostat sections through the entire right cerebellar half of a rat brain. For volume estimation, every fifth section[2] of the complete series was selected (systematic aspect of sampling), and section no. 1 was randomly selected from the first 5 sections[3] of the cerebellar half (random aspect of sampling). The distance between the upper surfaces of the selected sections was 250 µm (5 × 50 µm). Note that the volume estimate is independent of

---

[1] Cavalieri, Geometria Indivisibilibus Continuorum, Typis Clementis Ferronij, Bononiae, 1635 (reprinted as Geometria Degli Indivisibili, Unione Tipografico-Editrice Torinese, Torino, 1966).
[2] The choice of every fifth section was made based on results of a pilot study.
[3] A new random number between 1 and 5 would then be chosen for each new animal.

shrinkage of the section thickness that may have occurred after sectioning.[1]

Then, the cross-sectional area of the region of interest is measured on all sections containing the region of interest. You can make this measurement using either point counting[2] or by tracing the boundaries of the region of interest on video images displayed on a computer.[3]

In our example, the region of interest was the granule cell layer in the right cerebellar half of the rat brain shown in Figure 1. Because of its complex shape, it would have been tedious and time consuming to trace the boundaries of the granule cell layer on all selected sections. Instead a rectangular lattice with uniform distances between the lines in directions X and Y of 450 µm was randomly placed on each selected section (as illustrated in Figure 1, part B), and the points formed by the intersections of the lattice and the cerebellar granule cell layer were counted. The following results were obtained: 1 intersection (point) on section no. 1, and 8 (no. 2), 33 (no. 3), 38 (no. 4), 51 (no. 5), 33 (no. 6), 38 (no. 7), 40 (no. 8), 40 (no. 9), 45 (no. 10), 32 (no. 11) and 11 (no. 12) intersections on the other sections, respectively.

The product of the sum of all cross-sectional areas of the region of interest and the uniform distance between the sections gives an unbiased estimate of the volume of the region of interest. In cases when point counting is used, you can obtain an estimate of the cross-sectional area under study by multiplying the sum of counted points with the uniform distances between the lines in directions X and Y.

In our example, each counted intersection of the lattice and the cerebellar granule cell layer represented an area of 450 × 450 µm or 0.2025 mm^2. Accordingly, the following estimates were obtained for the profile areas of the cerebellar granule cell layer: 0.203 mm^2 for section no. 1, and 1.620 mm^2 (no. 2), 6.683 mm^2 (no. 3), 7.695 mm^2 (no. 4), 10.328 mm^2 (no. 5), 6.683 mm^2 (no. 6), 7.695 mm^2 (no. 7), 8.100 mm^2 (no. 8), 8.100 mm^2 (no. 9), 9.113 mm^2 (no. 10), 6.480 mm^2 (no. 11) and 2.228 mm^2 (no. 12) for the other sections, respectively. The sum of all estimated profile areas of the granule cell layer was

---

[1] It is possible for volumes of some biological regions to be affected by perfusion, dissection, and presectioning preparation. In these cases volume estimates derived using the Cavalieri estimator will not reflect the original *in vivo* volumes.
[2] Gundersen and Jensen, J Microsc 1987;147:229.
[3] Glaser and Glaser, J Chem Neuroanat 2000,20:115.

74.925 mm², leading to an estimate of the volume of the granule cell layer in the right cerebellar half of the rat brain shown in Figure 1, part A of 18.7 mm³.[1] Note that this estimate did not take any shrinkage of the tissue during fixation or histological processing into account.

It has been noted in the literature that Cavalieri estimates could become biased due to a phenomenon known as overprojection when thick sections are investigated under the microscope.[2] To compensate for the possibility of overprojection resulting in overestimation of the volume of the region of interest it has been proposed to disregard the section with the largest cross-sectional area.[3] However, overestimation due to overprojection in this context can be almost eliminated by performing the microscopic analyses using an objective providing a narrow depth of focus, using proper Köhler illumination.

The number of points to be counted (or, more importantly, the appropriate grid spacing to use) for estimating the volume of a region of interest with the Cavalieri estimator is addressed in the section *How does consideration about variability affect the implementation of design-based stereological estimators?* on page 50 of this book.

How can I quantify the total number of cells in my region of interest?

In general, investigations on the number of cells in a region of interest are complicated by the fact that it is usually not possible to physically isolate the cells. Accordingly, it is necessary to cut the tissue into sections, and to find the number of cells by inspecting the sections. This, however, leads to the problem that cutting a given tissue volume into sections also results in cutting the cells in this tissue. Obviously, the number of cell fragments in the sections differs from the original number of cells in the tissue, and as a consequence, estimates of cell numbers based solely on counts of cell fragments in sections are biased.[4] Figure 3 and Figure 4 show a solution to this problem.

---

[1] The volume of the molecular layer in the right cerebellar half of the rat brain shown in Figure 1, part A was 22.8 μm³ and the volume of the white matter (including the cerebellar nuclei) was 11.9 mm³.
[2] Uylings et al., J Neurosci Methods 1986;18:19.
[3] Gundersen and Jensen, J Microsc 1987;147:229.
[4] Cell population estimates based on cell profile counts will always overestimate the cell population, by an indeterminate amount, unless issues of cell size, shape, and orientation are somehow also taken into account, which is generally difficult or impossible to do. The amount of bias will not necessarily be equal between groups, so even these biased results cannot be safely compared without additional insight into changes in size, shape, and orientation.

Design-based stereology provides several methods to perform this quantification. The first method is called the Optical Fractionator. It uses thick sections and estimates the total number of cells from the number of cells sampled with a SRS set of unbiased virtual counting spaces covering the entire region of interest with uniform distance between unbiased virtual counting spaces in directions X, Y and Z.[1] Another method calculates the mean cell density within the unbiased virtual counting spaces by dividing the number of cells counted within all counting spaces by the number of investigated counting spaces and their uniform volume. When multiplying this average cell density with the volume of the investigated region of interest (estimated with the Cavalieri estimator), one obtains an unbiased estimate of the total number of cells in the region of interest (shown in Figure 1, part B). This principle has been named the $N_V \times V_{Ref}$ method.[2] Because no density measurements are required for estimation of number with the Optical Fractionator, it is easier to implement. In addition, the Optical Fractionator is immune to shrinkage of the tissue due to histological processing. Additionally, it is easier to estimate the precision of the number estimate (the CE) generated using the Optical Fractionator, making it the method of choice.

The counting criteria for both methods are as follows:

- Decide on a unique point (the *characteristic point*) that can be identified for each cell. This can be the top of the cell, i.e., the first part of the cell to come into focus while focusing down through the tissue, or perhaps the top of the nucleus, or even the nucleolus if the cells being examined have only one nucleolus per cell.

- The unique point must be visible for each cell to be counted.

- At each investigated microscopic field, an unbiased counting frame consisting of two exclusion lines and two inclusion lines[3] is placed on the section.

- Select a cell which is a potential candidate for counting and focus on its selected 'point' (e.g., 'top', nucleolus, etc.) and follow the rules below for counting:

---

[1] A related method, the Physical Fractionator, also known as the Physical Disector, utilizes a similar technique with (normally adjacent) pairs of thin sections. It also uses systematic random sampling for selection of section pairs, as well as for sampling sites on the sections.

[2] This method combines an estimate of the volume of the reference space of a region of interest (V_{Ref}) with an estimate of the mean cell density in this region (N_V).

[3] Gundersen, J Microsc 1977;111:219.

1. Any point that touches the exclusion line, crosses the exclusion line, or is outside of the counting frame is not counted.

2. For any point that is still a valid candidate and that is a) inside the counting frame, b) touching an inclusion line from either inside or outside the counting frame, or c) crossing the an inclusion line, remains a candidate to be counted (Figure 3).

3. If an object touches or crosses both an inclusion line and an exclusion line, it is not counted (as per rule #1 above).

4. Restrict your analysis to a virtual space within the section thickness (Figure 4 and Figure 5). For any cell that is a valid candidate, count it only if its unique point- in addition to its position relative to the unbiased counting frame (as per rules 1, 2, and 3) - comes into focus within this virtual space. In other words, a cell is counted only if its unique point[1] is found within an unbiased virtual counting space.[2]

The top of the perikaryon (or nucleus or nucleolus, respectively) of a cell is a unique point that is located at only one unique position in space. The cell counts carried out with unbiased virtual counting spaces are unbiased in that they are not influenced by the size, shape, spatial orientation, and spatial distribution of the cells under study.

---

[1] König et al., J Microsc 1991;161:405.
[2] The term 'unbiased virtual counting space' was introduced in the literature by Schmitz (Anat Embryol 1998;198:371) and Schmitz and Hof (J Chem Neuroanat 2000;20:93). Other related terms are 'Optical Disector' (West et al., Anat Rec 1991;231:482; West, Neurobiol Aging 1993;14:275; West and Slomianka, Trends Neurosci 2001;24:374) and 'counting box' (Williams and Rakic, J Comp Neurol 1988; 278:344).

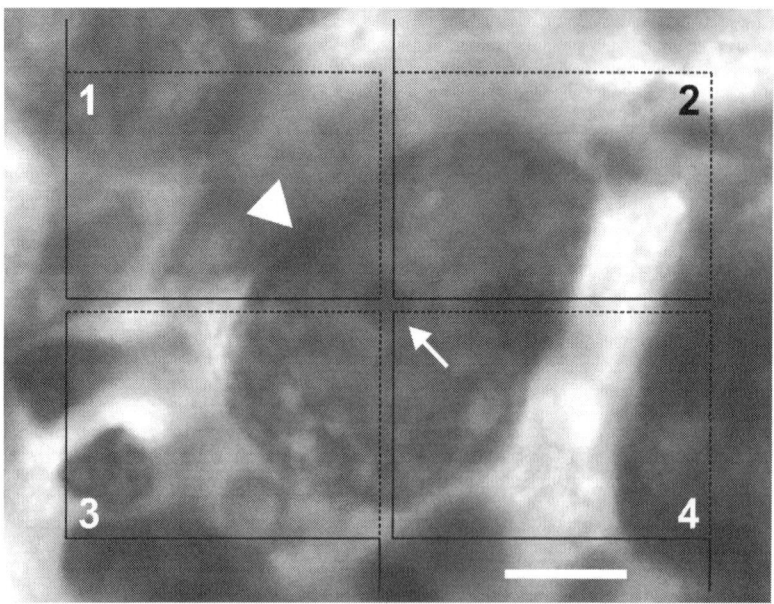

Figure 3. Mode of operation of the two-dimensional unbiased counting frame (UCF), illustrated for the cerebellar Purkinje cell (arrowhead) marked by an asterisk in Figure 1, part D shown here at higher magnification (rat; 50 μm thick, Nissl-stained coronal cryostat section). Four UCFs are shown, positioned closely together. Each UCF consists of two *exclusion lines* (solid lines in this example) and two *inclusion lines* (dashed lines in this example).[1] *Objects of interest* (i.e., cells, nuclei or nucleoli, etc.) are counted provided they are found entirely within an UCF or hit at least one of the inclusion lines of an UCF but not any of the exclusion lines of the same UCF. Accordingly, the Purkinje cell shown in this example is counted in UCF no. 3 (and only in this UCF), because it hits at least one of the exclusion lines of the UCFs no. 1, 2 and 4 each. If the UCFs no. 1, 2 and 4 would have been positioned at larger distance to UCF no. 3 (as shown in Figure 1, part C) the criteria for counting would have been the same. If one considers the nucleoli of cerebellar Purkinje cells as objects of interest (the arrow), the nucleolus of the cell shown in the example would also be counted in UCF no. 3. This illustrates how the use of UCFs makes it possible to count objects of interest independent of the objects' size.[2] Note that in design-based stereology also inclusion / exclusion criteria in the third dimension (i.e., the section thickness in histology) have to be considered. This is shown in Figure 4. Scale bar represents 10 μm.

---

[1] For sake of clarity, the lower parts of the exclusion lines of UCFs no. 1 and 2 were omitted, as well as the upper parts of the exclusion lines of UCFs no. 3 and 4.

[2] One can even consider punctate objects such as the midpoint of the nucleolus of cerebellar Purkinje cells as *objects of interest*. In this case one counts so-called *characteristic points* (König et al.; J Microsc 1991;161:405). Provided the inclusion and exclusion lines of the UCFs being infinitesimally thin, counting of punctate characteristic points with UCFs follows the same rules as described above.

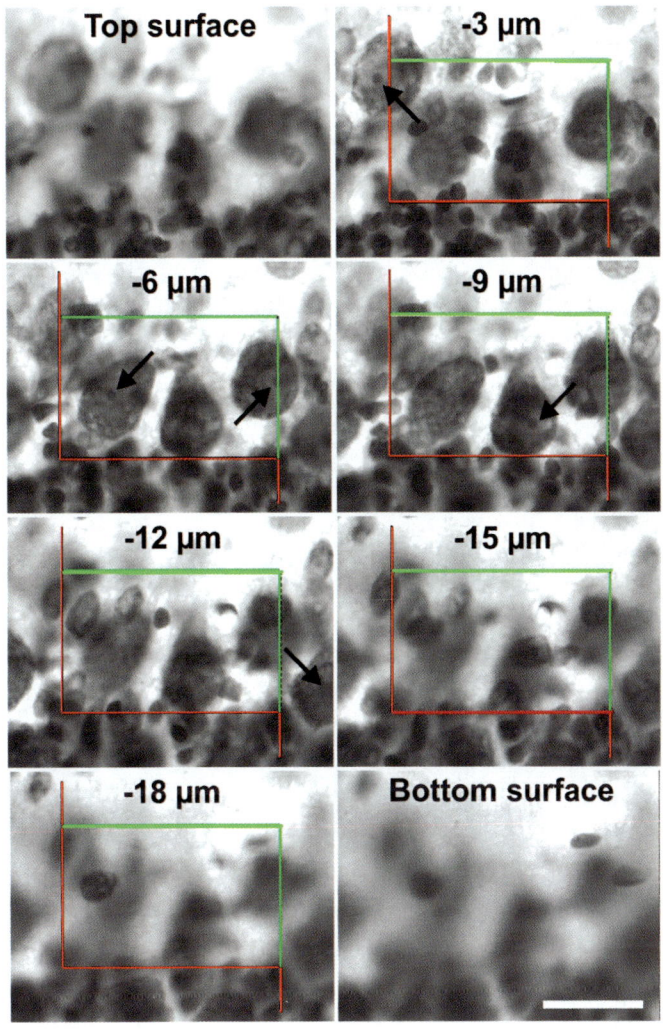

Figure 4 Counting cells with three-dimensional *unbiased virtual counting spaces* (or *Optical Disectors*, respectively) in thick, Nissl-stained sections, illustrated for the microscopic field from the rat cerebellum shown in Figure 1, part D (50 μm thick coronal cryostat section). The figure shows the selected microscopic field at the top surface of the section, at six consecutive focal planes below the top surface with a distance of 3 μm between the focal planes, and at the bottom surface of the section that was found 20 μm below the top surface.[1] Between -3

---

[1] Note that a cryostat section from a perfusion-fixed brain may show shrinkage in the section thickness by up to 60% during histological processing. Cryostat sections from snap-frozen brains can even show shrinkage in the section thickness by up to 90% during histological processing (see Schmitz et al.; J Chem Neuroanat 2000; 20:21).

μm and -18 μm the microscopic field is superimposed by an unbiased counting frame (UCF); explained in detail in Figure 3). This UCF served as basis for an *unbiased virtual counting space* (or *Optical Disector*, respectively) with base area of the UCF and a height of 15 μm in this example. *Objects of interest* (i.e., cells, nuclei or nucleoli, etc.) are counted provided (i) they are found entirely within an UCF or hit at least one of the inclusion lines (green lines in this example) of an UCF but not any of the exclusion lines (red lines in this example) of the same UCF, and (ii) they come into focus within the unbiased virtual counting space for which the corresponding UCF serves as basis. In the illustrated example, these criteria were applied to the nucleoli of the cerebellar Purkinje cells as objects of interest. As shown, three nucleoli of Purkinje cells (marked with arrows at -6 μm and -9 μm) fulfilled both criteria and were therefore counted. Two other nucleoli of Purkinje cells came into focus at -3 μm and -12 μm, respectively (also marked with arrows), i.e., within the range of the section thickness covered by the height of the unbiased virtual counting space. However, both nucleoli were not found entirely within the UCF or at least hitting one of its inclusion lines and were therefore not counted. Provided the nuclei or the perikarya of the Purkinje cells would have been selected as objects of interest in this example, also, the cells found at -6 μm and -9 μm would have been counted. Scale bar represents 40 μm.

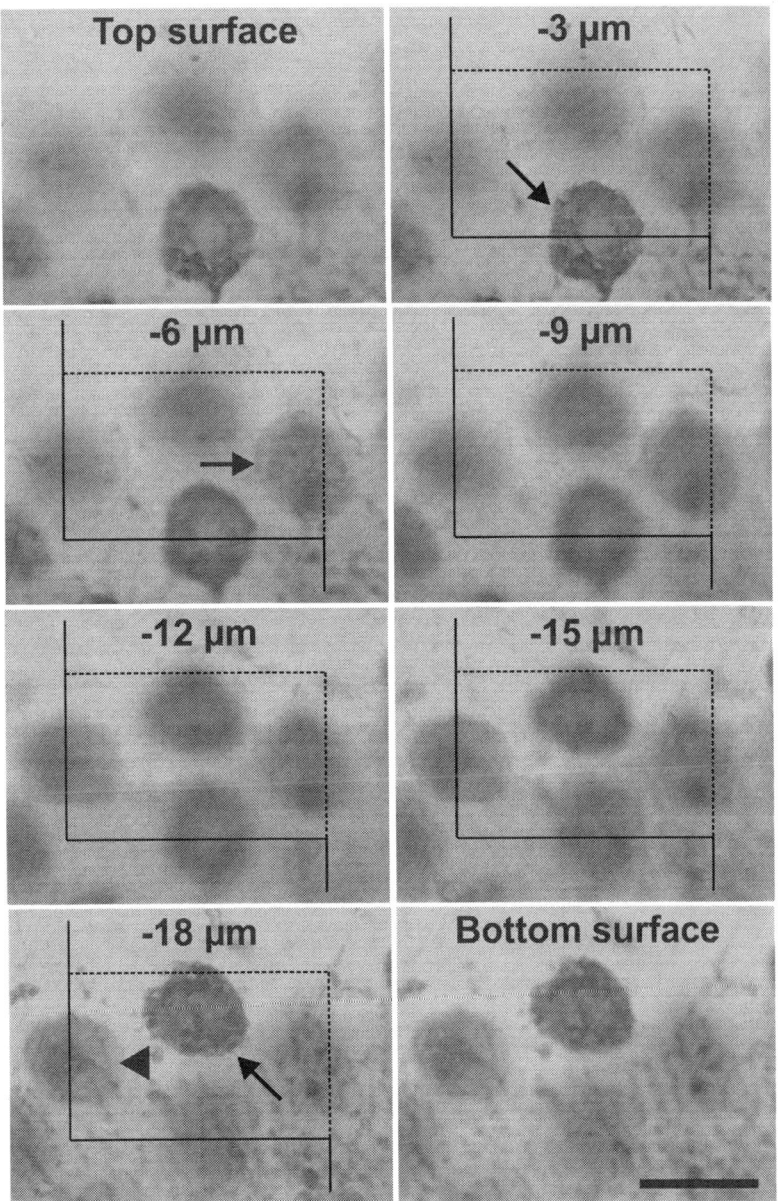

Figure 5. Counting cells with three-dimensional unbiased virtual counting spaces (or Optical Disectors, respectively) in thick sections processed with immunohistochemistry, illustrated for a microscopic field from an adjacent section to the one shown in Figure 1, part D as well as in Figure 3 and Figure 4 from the same rat cerebellum (50 μm thick coronal cryostat section).

The section was processed with immunohistochemistry for the detection of calbindin D-28k, a marker of cerebellar Purkinje cells.[1] The figure shows the selected microscopic field at the top surface of the section, at six consecutive focal planes below the top surface with a distance of 3 μm between the focal planes, and at the bottom surface of the section that was found 20 μm below the top surface. Between -3 μm and -18 μm the microscopic field is superimposed by an unbiased counting frame (UCF); explained in detail in Figure 3). This UCF served as basis for an *unbiased virtual counting space* (or Optical Disector, respectively) with base area of the UCF and a height of 15 μm in this example. *Objects of interest* (i.e., cells, nuclei or nucleoli, etc.) are counted provided (i) they are found entirely within an UCF or hit at least one of the inclusion lines (dashed lines in this example) of an UCF but not any of the exclusion lines (solid lines in this example) of the same UCF, and (ii) they come into focus within the unbiased virtual counting space for which the corresponding UCF serves as basis. In the illustrated example, these criteria were applied to the perikarya of the cerebellar Purkinje cells as objects of interest. As shown, a Purkinje cell was found at -3 μm (marked with an arrow). However, the same cell already came into focus at the top surface of the section and, thus, not within the unbiased virtual counting space. Furthermore, this cell hit the exclusion line of the UCF and was therefore not counted. Two other Purkinje cells came into focus at -18 μm, i.e., within the range of the section thickness covered by the height of the unbiased virtual counting space. One of them (marked with an arrow) hit one of the inclusion lines of the UCF and was therefore counted, whereas the other cell (marked with an arrowhead) hit one of the exclusion lines of the UCF and was therefore not counted. A fourth Purkinje cell came into focus at -6 μm, hit one of the inclusion lines of the UCF, and was therefore also counted. Note that the perikaryon of this cell appeared somewhat blurry, which was most probably due to incomplete penetration of either the primary or the secondary antibody through the section thickness. This is a common problem in immunohistochemistry on thick sections and has been addressed in the stereological literature.[2] In certain combinations of tissue, section thickness, and antibodies, the immunohistochemical reaction might not work in the center of the section but is restricted to the upper and lower parts of the section thickness. This must be evaluated by careful microscopic inspection of the sections prior to stereological analysis, and the height of the unbiased virtual counting spaces and their position within the section thickness must be selected accordingly. Disregarding this problem (i.e., placing unbiased virtual counting spaces in the center of the section thickness despite improper immunohistochemical detection of the antigen of interest at this place) can cause serious bias in stereological investigations. Scale bar represents 40 μm.

There are two important and common potential sources of bias in cell counts with unbiased virtual counting spaces. The first one may arise from loss of cells (or cell perikarya, nuclei, or nucleoli, respectively) at the upper or the lower surface of sections when hit by the knife during sectioning of the tissue.

---

[1] Free-floating sections. Primary antibody: mouse monoclonal anti-calbindin D-28k (1:30,000; C9848; Sigma, St. Louis, MO, USA). Secondary antibody: biotinylated anti-mouse-IgG (1:600; Vector Laboratories, Burlingame, CA, USA), followed by reaction with preformed avidin-biotin peroxidase complex (ABC Vectastain; Vector Laboratories). Visualization of immunoreactivity with 3,3'-diaminobenzidine (DAB, 0.015%) in the presence of 0.007% hydrogen peroxide.
[2] Jinno et al.; Brain Res 1998; 814:55.

This phenomenon, known as lost caps[1] or plucked cells, occurs if the top of cells (or cell perikarya or nuclei or nucleoli, respectively), or the cells themselves, could become lost and could therefore not come into focus when unbiased virtual counting spaces are placed close to the upper and lower surface of the sections. Accordingly, you would underestimate the number of cells in the investigated tissue with unbiased virtual counting spaces. Although it has been shown that lost caps can be prevented using adequate histological techniques,[2] placing the upper surface of the unbiased virtual counting spaces some distance below the upper surface of the sections (i.e., introducing a 'guard zone' larger than zero; see Figure 4), as well as placing the lower surface of the unbiased virtual counting spaces some distance above the lower surface of the sections, will help to prevent potential bias in estimates of total numbers of cells. This will also help ameliorate problems stemming from uneven or wavy surfaces of the sections.

A second potential source of bias is incomplete staining of the tissue, particularly in the middle of the section thickness. This problem, which may result in underestimation of the number of cells in a given tissue volume with unbiased virtual counting spaces could arise particularly when using immunohistochemical processing (as shown in Figure 5). In general, it is always recommended first to carry out a pilot study on the sections prior to stereological analysis, to determine the thickness of the sections showing adequate staining. Then, the height of the unbiased virtual counting spaces should be adjusted appropriately.

In most experimental settings it is not possible to count all cells in the region of interest. Rather, you have to select a proper sample of microscopic fields to be investigated, and derive an estimated total number of cells in the region of interest from the number of cells in the sample along with the sampling probability. Figure 1, part C shows the solution to this problem in design-based stereology (for a single section). Unbiased virtual counting spaces are placed in an SRS manner within a series of systematically and randomly sampled sections throughout the region of interest and cells are counted according to the criteria discussed above at the selected microscopic fields. The easiest way to achieve this is to place a rectangular lattice on the surface of the sections, determining the positions of the unbiased virtual counting spaces which partially or completely overlap the region of interest (as shown in Figure 1, part C).

To obtain a number estimate with the Optical Fractionator when cells are counted with unbiased virtual counting spaces, multiply the number of

---

[1] Andersen and Gundersen, J Microsc 1999;196:69.
[2] Schmitz et al., J Chem Neuroanat 2000;20:21.

cells counted within all counting spaces with the reciprocal value of the sampling fraction (i.e., sampling probability). The latter depends on the following three fractions: 1) the number of investigated sections compared to the total number of sections (the section sampling fraction or *ssf*), 2) the base area of the unbiased virtual counting spaces compared to the product of the side lengths of the rectangular lattice used for placing the UCFs within the sections (the area sampling fraction or *asf*), and 3) the height of the unbiased virtual counting spaces compared to the average section thickness after histological processing (the thickness sampling fraction or *tsf*).

In our example, we wanted to estimate the total number of Purkinje cells in the right cerebellar half of the rat brain shown in Figure 1. To this end, we counted Purkinje cells in the SRS series of sections shown in Figure 1 with unbiased virtual counting spaces. The base area of the unbiased virtual counting spaces (B) was 9,375 µm^2 (125 µm x 75 µm), the height of the unbiased virtual counting spaces (H) was 7 µm, and the top of the unbiased virtual counting spaces was placed 3 µm below the upper surface of the sections that had an average thickness (t) of 25.3 µm after histological processing.[1] The distance between the unbiased virtual counting spaces in directions X and Y (D) was 300 µm. This sampling scheme resulted in a total number of 1,670 unbiased virtual counting spaces hitting the cerebellar half, and a number of 898 Purkinje cells that were counted. Accordingly, the sampling probability was calculated as follows:

- ssf = 1/5 = 0.2
- asf = B/D^2 = 9,375 µm^2 / (300 µm × 300 µm) = 0.104
- tsf = H / t = 7 µm / 25.3 µm = 0.277
- Sampling probability^{-1} = (ssf × asf × tsf)$^{-1}$ = 173.5

Accordingly, the estimated total number of Purkinje cells in the right cerebellar half of the rat brain shown in Figure 1 was 898 × 173.5 = 155,803.[2]

The number of cells to be counted for estimating the number of cells within a region of interest is addressed in the section *How does consideration about variability affect the implementation of design-based stereological estimators?* on page 50 of this book.

---

[1] Note that from a theoretical point of view, this is: a) a relatively thin top guard zone, b) a sampling space height that will require an oil objective and ideally an oil condenser, plus c) a sufficiently large bottom guard zone.

[2] Estimated total number of Purkinje cells in the entire cerebellum of rats have been mainly reported between 190,000 and 340,000 (for details see Schmitz et al., J Chem Neuroanat 2000;20:21).

How can I measure the size of the cells in my region of interest?

There are several methods you can use to estimate mean perikaryal or nuclear volumes of cells. To estimate number-weighted mean volumes (in which all cells in the region of interest have the same probability of being selected for investigation) use the Nucleator[1] (Figure 6, part A), the Rotator[2] (Figure 6, part B) or the Optical Rotator[3] (not shown). These can be implemented in conjunction with a Physical or Optical Fractionator sampling scheme in order to achieve efficient systematic random sampling. When implemented with the Optical Fractionator, cell population estimates and cell volume estimates can be achieved at the same time.[4] Alternatively, you can use Point Sampled Intercepts[5] (Figure 6, part C) to estimate volume-weighted mean volumes (in which the probability of a cell being selected for investigation depends on its individual size).

Estimates of mean perikaryal or nuclear volumes of cells with the Nucleator, the Rotator or with Point Sampled Intercepts have to be performed on isotropic uniform random or vertical sections.[6] In preparing both isotropic uniform random sections and vertical sections, however, you cannot select a certain plane of section that might be necessary to identify the region of interest. This problem does not occur when using the Optical Rotator, which works well on sections showing only a minimum shrinkage in the z-axis. Such sections can be prepared when, for example, tissue is embedded in methacrylate.[7] This procedure, however, is not suitable in many neuroscience applications because it prevents processing of tissue sections with immunohistochemistry. On the other hand, frozen or vibratome sections can show considerable shrinkage in the z-axis,[8] preventing the use of the Optical

---

[1] Gundersen; J Microsc 1988;151:3.
[2] Vedel Jensen and Gundersen; J Microsc 1993;170:35.
[3] Tandrup et al., J Microsc 1997;186:108.
[4] This requires isotropic or vertical sectioning.
[5] Vedel Jensen and Gundersen; J Microsc 1993;170:35.
[6] For the concept of isotropic uniform random sections see Miles and Davy (J Microsc 1976;107:211), and of vertical sections Baddeley et al., J Microsc 1986;142:259). Protocols to prepare such sections from a region of interest have been provided in the literature by Baddeley et al., (1986), Mattfeld et al. (J Microsc 1990;159:301), Nyengaard and Gundersen (J Microsc 1992;165:427) and Schmitz et al. (J Neurosci Methods 1999;88:71).
[7] For the preparation of methacrylate sections from brain tissue see Sousa et al. (Neuroscience 1999;89:1079) or Lukyanov et al. (Alcohol 2000;20:139).
[8] For details, see Messina et al. (J Neurosci Methods 2000;97:133), Schmitz et al. (J Chem Neuroanat 2000;20:21) and Dorph-Petersen et al. (J Microsc 2001;204:232).

Rotator in most applications. This problem has to be solved individually in each application.[1]

---

[1] Note that it could be shown for hippocampus and cerebellum of the mouse brain that there are no differences in estimates of mean nuclear volumes obtained with the Nucleator, the Rotator and with Point Sampled Intercepts on either isotropic uniform random sections or sagittal sections (Schmitz et al., J Neurosci Methods 1999;88:71).

Figure 6. Estimating the size of objects of interest with design-based stereological methods, illustrated for the perikaryon of the cerebellar Purkinje cell marked by an asterisk in Figure 1, part D shown here at higher magnification (50 μm thick Nissl-stained coronal cryostat section of rat cerebellum). In all examples, the Purkinje cell has been selected by a UCF in the focus plane of the midpoint of the nucleolus (m). **A**, The Nucleator.[1] (i) Two mutually orthogonal lines are generated through the midpoint of the nucleolus, and the intersections of these lines and the boundary of the perikaryon are identified (triangles). (ii) The distances between these intersections and the midpoint of the nucleolus (l_m) are measured. (iii) The perikaryal volume of the cell can be calculated from the average of the third powers of these measurements. **B**, The Rotator[2], also called The *Planar Rotator*.) (i) An axis is generated along the shortest aspect of the perikaryon (line "1"). (ii) The 'height' of the perikaryon (i.e., the extension of the perikaryon along line 1) is measured (h). (iii) The top and the bottom position of the perikaryon in respect of line 1 are marked with lines placed orthogonal to line 1 (lines "2"). (iv) Three parallel test lines are generated parallel to lines 2 (lines "3"). The position of these lines in respect of line 1 is uniformly random, i.e., the position of the first line is randomly chosen in an interval of h/3, with the distance between the lines being h/3. (v) The intersections of lines 3 and the boundary of the perikaryon are

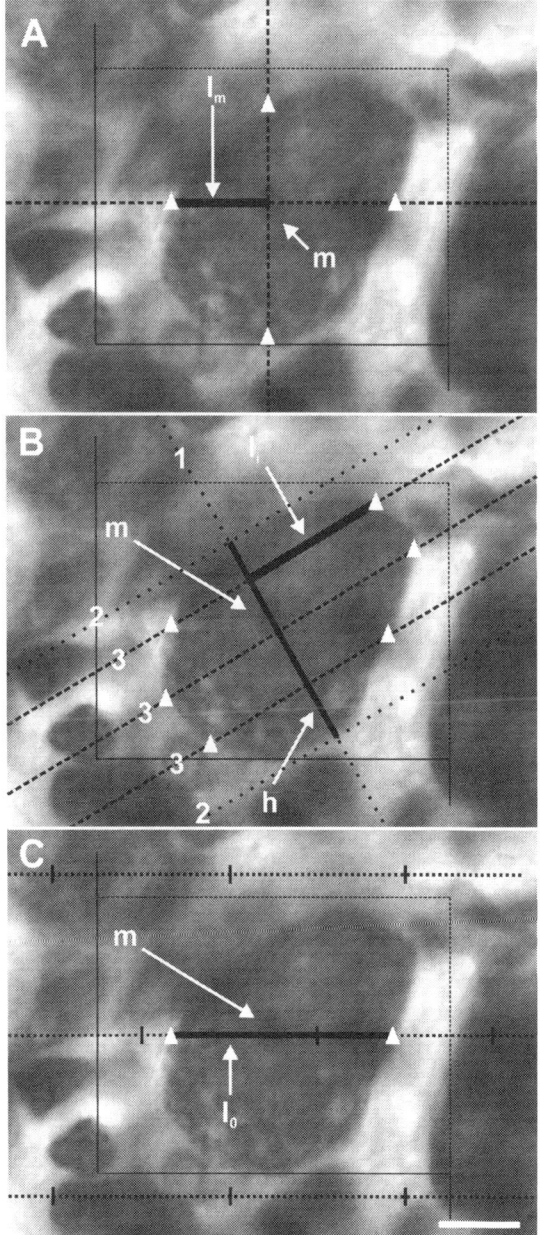

[1] Gundersen; J Microsc 1988;151:3.
[2] Vedel Jensen and Gundersen; J Microsc 1993;170:35.

identified (triangles). (vi) The distances between these intersections and line 1 are measured (l), and the perikaryal volume of the cell can be calculated from these measurements. **C**, Point Sampled Intercepts.[1] (i) An array of parallel test lines with intersections is randomly placed on the section. (ii) Provided an intersection of a test line hits the perikaryon, the length of the intercept of the corresponding test line between the boundaries of the perikaryon is measured (l_o). The perikaryal volume of the Purkinje cell can then be calculated based on these measurements. Note that the application of the Nucleator and the Rotator leads to estimates of *number-weighted mean local volumes*, i.e., each object of interest has the same probability of being selected for analysis. In contrast, use of Point Sampled Intercepts results in *volume-weighted mean local volumes*, i.e., the probability of an object of interest being selected for investigation depends on its individual size. Scale bar represents 10 μm.

The number of cells that must be sampled in a region of interest to be investigated when estimating their mean perikaryal or nuclear volume is addressed in the section *How does consideration about variability affect the implementation of design-based stereological estimators?* on page 50 of this book.

How can I determine the length density and total length of capillaries or cell processes in my region of interest?

Design-based stereology offers several methods in this regard.[2] One method is to estimate the length of tubular objects such as capillaries or cell processes with systematically spaced straight lines, or a series linked semicircles (Merz), provided thin isotropic uniform random sections through the region of interest are available.[3] With vertical sections, systematically spaced sine-weighted curves (cycloids) can be used.

A second method is to quantify tubular objects with isotropic virtual planes.[4] In this case, the tubular objects under study contained in thick (3D) sections are investigated with software-randomized isotropic virtual planes in volume probes in systematically sampled microscopic fields. A disadvantage of this technique is the fact that the analysis has to be carried out on virtual planes within the thick sections, the counting rules for which can be tedious and cumbersome.

Another solution is the Space Balls method[5] as shown in Figure 7. In a first step, an SRS set of sampling sites encompassing the 3D region of interest is

---

[1] Gundersen and Jensen; J Microsc 1985;138:127.
[2] Space Balls, Isotropic Virtual Planes, Petrimetrics, IUR Optical Fractionator, L-Cycloid Optical Fractionator, Cycloids for Lv, Merz, and Weibel amongst others.
[3] For details, see Calhoun and Mouton, J Chem Neuroanat 2000;20:61.
[4] Larsen et al., J Microsc 191:238.
[5] Calhoun and Mouton, J Chem Neuroanat 2000;21:61; Mouton et al., J Microsc 2002;206:54.

generated. Then, virtual spheres (or hemispheres[1]) are placed within the sections at all microscopic sampling sites, and the intersections between the spheres (or hemispheres) and the tubular objects under study are counted. Note that the mounted tissue thickness must be greater than the diameter (for spheres) or radius (for hemispheres) of the Space Balls. From this data you can obtain both the length density of the linear biological structures as well as the total length of the tubular objects.

To demonstrate the use of the Space Balls method, the capillary length density in the lateral nucleus of the amygdala in the right hemisphere from a human postmortem brain was determined (Figure 7). Based on the analysis of 517 hemispheres with a radius of 30 μm[2], a capillary length density of 508 mm/mm^3 was found. The volume of the lateral nucleus of the amygdala nucleus was 384 mm^3 (estimated with the Cavalieri estimator.[3] Accordingly, the total capillary length in this nucleus was 195 μm.

---

[1] Hemispheres are better suited to thinner tissue and in situations with a very high density of fibers.
[2] This required a final mounted section thickness of more than 30μm.
[3] In a recent design-based stereological study a mean volume of the lateral nucleus of the amygdala in the human brain of 452 ± 43 mm^3 (mean ± standard deviation) was reported. (Schumann and Amaral, J Comp Neurol 2005;491:320).

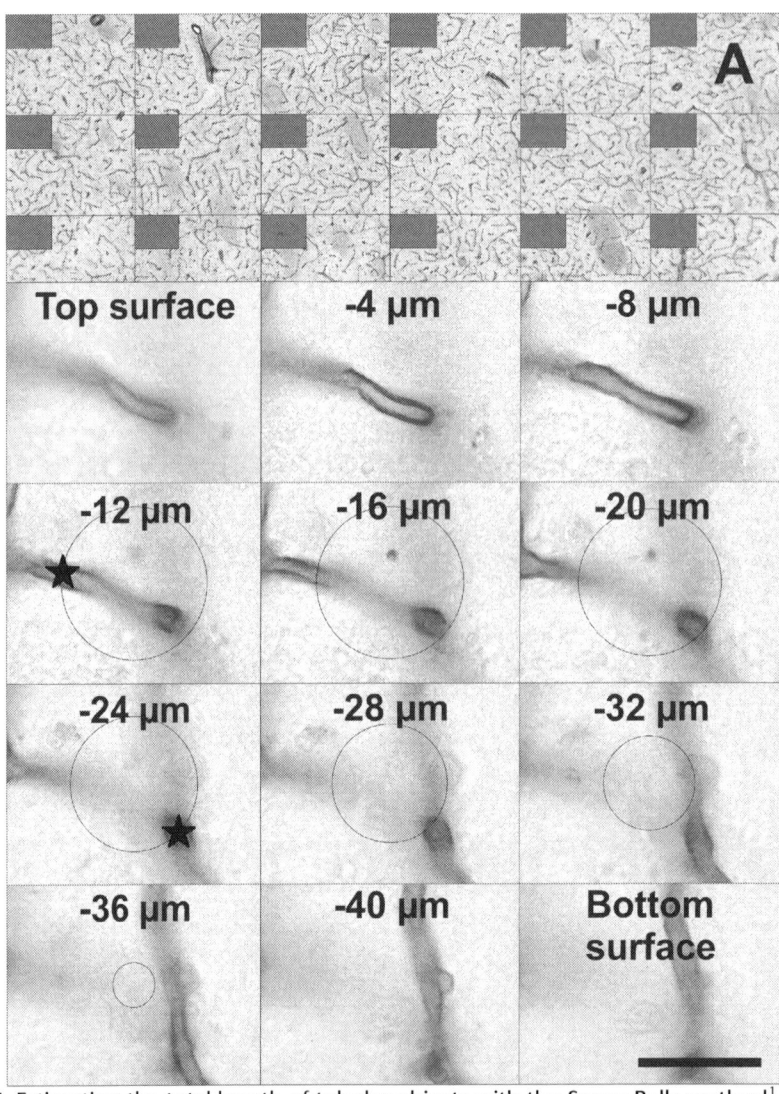

Figure 7. Estimating the total length of tubular objects with the Space Balls method[1], illustrated for collagen IV-immunoreactive capillaries in the lateral nucleus of the amygdala in the right hemisphere from a human postmortem brain (50 μm thick coronal cryostat section).[2] **A,** Low-power overview of capillaries, superimposed by a rectangular lattice with

---

[1] Calhoun and Mouton, J Chem Neuroanat 2000;21:61; Mouton et al., J Microsc 2002;206:54.
[2] 50-year-old male; cause of death: avalanche accident; interval between death and autopsy 23 hours; immersion-fixation in 10% formalin for 1.4 years prior to histological processing; free-floating sections. Primary antibody: mouse monoclonal anti-collagen IV (1:500; #C1926, clone COL-94; Sigma Aldrich, St. Louis, MO, USA). Secondary antibody: biotinylated anti-mouse IgG (1:200; Jackson ImmunoResearch, West Grove, PA, USA), followed by preformed avidin-biotin peroxidase complex (ABC

uniform distance between the lines in directions X and Y. This lattice defines the position of microscopic fields (indicated as gray rectangles), at which the section is inspected at higher magnification to perform Space Balls analyses. The high-power photomicrographs show one of these microscopic fields at the top surface of the section, at ten consecutive focal planes below the top surface with a distance of 4 μm between the focal planes, and at the bottom surface of the section that was found 43 μm below the top surface.[1] Between -12 μm and -36 μm the microscopic field is superimposed by intersections of a hemisphere (a semi- Space Ball) with the focal plane at the investigated focal depth, illustrated as circles. The (semi-) Space Ball was centered at a depth of -12 μm and had a radius of 25 μm. Intersections between the (semi-) Space Ball and the capillaries in focus were found at -12 μm, and -24 μm (asterisks). The average capillary length density within the region of interest can be calculated from the total number of intersections and the number and size of the (semi-) space balls. The total capillary length within the region of interest can then be obtained by combining the average capillary length density and the volume of the region of interest obtained with the Cavalieri estimator. Scale bar represents 700 μm in **A**, and 40 μm in the high-power photomicrographs.

It should be mentioned that estimates of length of tubular objects obtained with Space Balls are not unbiased in a strict sense. One source of potential bias is differential shrinkage of the tissue sections in the Z dimension due to histological processing. As a result, length and length density estimates can be different if sections are cut in different planes, and the tubular objects are not isotropic. Another potential source of bias is the ratio of the diameter of the tubular objects and the diameter of the spheres.[2] For this reason, it is recommended that Space Balls be used primarily to estimate the length of relatively thin tubular objects. Practical solutions to minimize bias in stereological quantification of tubular objects are provided in the literature.[3]

The number of spheres (or hemispheres), and their size, to be applied for estimating the length density of tubular objects in a region of interest with the Space Balls method is addressed in the section *How does consideration about variability affect the implementation of design-based stereological estimators?* on page 50 of this book.

Vectastain; Vector Laboratories, Burlingame, CA, USA). Visualization of immunoreactivity with 3,3'-diaminobenzidine (DAB, 0.8 mg/ml) in the presence of 0.003% hydrogen peroxide.

[1] Note that the section in this example showed much less shrinkage in section thickness than the sections described in Figure 4 and Figure 5. This was achieved by not dehydrating the section but mounting it on gelatin-coated slides and coverslipping it with 80% glycerol in tris-buffered saline. In this way, the section can be stored and analyzed for approximately 6 months at room temperature before the DAB begins to fade out. To archive the section the coverslip can be removed after stereological analysis, and the section can be dehydrated and then be coverslipped with, e.g., DPX.

[2] Gundersen, J Microsc 2002;207:155.

[3] Ibid.

How can I determine the spatial distribution of cells in my region of interest?

An elegant answer to this question was found by adopting the Nearest-Neighbor Distance Distribution Function (NNDDF) analysis founded in theoretical statistics[1] for use in quantitative histology.[2]

An NNDDF analysis comprises and integrates an estimate of the volume of a region of interest (obtained with the Cavalieri estimator) together with an estimate of the total number of cells in this region (obtained with the Optical Fractionator). For each cell in an SRS sample, the identification of its nearest-neighbor (in 3D) and the distance to its nearest-neighbor cell is determined (shown in Figure 8). From this data, the cumulative relative frequency distribution of the nearest-neighbor distances is obtained and can be graphically compared to corresponding distribution functions obtained from computer simulations. Empirical Distribution Function (EDF) plots are used to model virtual regions of interest with the same quantitative characteristics (i.e., volume, total number of cells) as the investigated region of interest.

NNDDF analyses can be applied to determine whether cells in a region of interest exhibit spatial randomness, a clustered distribution, or a more dispersed distribution.

NNDDF analyses require a guarantee that for each cell in the SRS sample of cells, the nearest-neighbor can be found within the same tissue section. Otherwise, the true nearest-neighbor distance cannot be determined for this cell, and the analysis could become biased. This, however, implies that the distance between the focus plane within the section in which the cells (for which the nearest-neighbors shall be found) are sampled and the upper and lower surface of the section must exceed the maximum nearest-neighbor distance of the cells in the SRS sample. In other words, NNDDF analyses require the use of thick tissue sections (between 100 µm and 150 µm for the rodent brain, and up to 750 µm for the human brain), stained by a method which guarantees staining throughout the section thickness. Appropriate protocols to generate such tissue sections have been published.[3]

---

[1] Diggle, Statistical Analysis of Spatial Point Patterns. Academic Press, New York, 1983.
[2] Schmitz et al., Cereb Cortex 2002;12:954.
[3] Heinsen and Heinsen, J Histotechnol 1991;14:167; Heinsen et al., J Chem Neuroanat 2000;20:49; Schmitz et al., Cereb Cortex 2002;12: 954.

Figure 8. Analysis of the spatial distribution of biological objects of interest in a given brain region with the Nearest-Neighbor Distance Distribution Function (NNDDF) method, illustrated for neurons in the lateral nucleus of the amygdala in the right hemisphere from the same human postmortem brain described in Figure 7 (500 µm thick, gallocyanin-stained coronal section obtained from tissue embedded in gelatin, deeply frozen at -60°C and then serially sectioned using a cryomicrotome). NNDDF analyses start with systematic-random selection of objects of interest as shown in Figure 1, part C, i.e., by superimposing the investigated section with a rectangular lattice with uniform distance between the lines in directions X and Y (not shown here). This lattice defines the position of microscopic

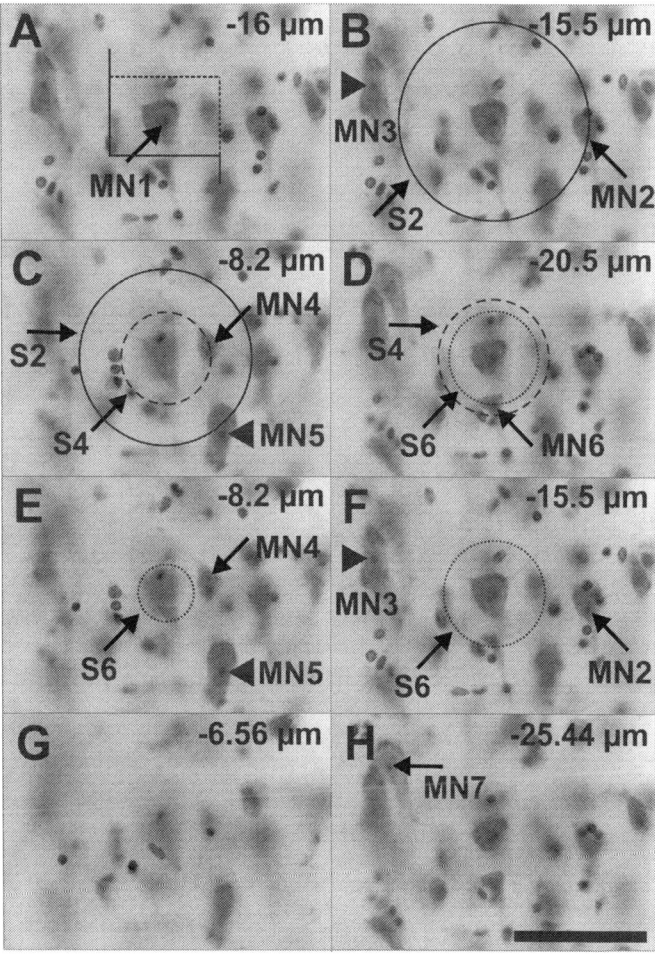

fields at which the section is inspected at higher magnification to perform NNDDF analyses. One of these microscopic fields is shown in this figure at high magnification and different focal planes. **A,** The midpoint of the nucleolus of a pyramidal neuron was selected by means of an unbiased counting frame at a depth of 16 µm below the top surface of the section (MN1). In the following steps, the nearest neighbor of this neuron (neuron #1) was determined. **B,** The midpoint of the nucleolus of another pyramidal neuron was found at a depth of 15.5 µm below the top surface of the section (MN2). It was hypothesized that this neuron (neuron #2) was the nearest neighbor of neuron #1. To test this hypothesis a sphere was constructed (S2[1]) that was centered on MN1 and had the three-dimensional distance

---

[1] The name S2 refers to the fact that this sphere was defined by the presence of neuron #2.

between MN1 and MN2 as radius.[1] The closed circle in B represents the plane intersection of S2 at the investigated focal depth. At the same focal depth, the midpoint of the nucleolus of a third pyramidal neuron was found (MN3). However, this neuron could not be the nearest neighbor of neuron #1 because MN3 was lying outside the circle representing S2 at the investigated focal depth.[2] **C**, At a depth of 8.2 μm below the top surface of the section, the midpoint of the nucleolus of a fourth pyramidal neuron was found (MN4). MN4 was lying inside the closed circle representing S2 at the investigated focal depth.[3] Accordingly, neuron #2 could not be the nearest neighbor of neuron #1.[4] To test the hypothesis that neuron #4 was the nearest-neighbor of neuron #1 another sphere was constructed (S4) that was centered on MN1 and had the three-dimensional distance between MN1 and MN4 as radius.[5] The dashed circle in C represents the plane intersection of S4 at the investigated focal depth. At the same focal depth, the midpoint of the nucleolus of a fifth pyramidal neuron was found (MN5). However, this neuron could not be the nearest neighbor of neuron #1 because MN5 was lying outside the circle representing S4 at the investigated focal depth. **D**, At a depth of 20.5 μm below the top surface of the section, the midpoint of the nucleolus of a sixth pyramidal neuron was found (MN6). MN6 was lying inside the dashed circle representing S4 at the investigated focal depth.[6] Accordingly, neuron #4 could not be the nearest neighbor of neuron #1. To test the hypothesis that neuron #6 was the nearest neighbor of neuron #1 a third sphere was constructed (S6) that was centered on MN1 and had the three-dimensional distance between MN1 and MN6 as radius.[7] The dotted circle in D represents the plane intersection of S6 at the investigated focal depth. **E**, By focusing back to a depth of 8.2 μm below the top surface of the section, it was found that both MN4 and MN5 were lying outside the dotted circle representing S6 at the investigated focal depth.[8] This confirmed the hypothesis that the neurons #4 and #5 were not the nearest neighbor of neuron #1. **F**, By focusing back to a depth of 15.5 μm below the top surface of the section, it was found that both MN2 and MN3 were lying outside the dotted circle representing S6 at the investigated focal depth.[9] This confirmed the hypothesis that the neurons #2 and #3 were not the nearest-neighbor of neuron #1. **G** and **H**, The focal planes 6.56 μm and 25.44 μm

---

[1] The three-dimensional distance between two points is defined as
$$D_{XYZ} = \sqrt{(x_1 - x_2)^2 + (y_1 - y_2)^2 + (z_1 - z_2)^2}$$
with x_1, y_1 and z_1 being the XYZ coordinates of the first point, and x_2, y_2 and z_2 being the XYZ coordinates of the second point. The three-dimensional distance between two points can also be calculated from the two-dimensional (XY) distance (D_{XY}) between the points and the absolute difference between the Z coordinates of the points (D_Z) as follows:
$$D_{XYZ} = \sqrt{D_{XY}^2 + D_Z^2}$$
In the example shown in B, D_{XY} between MN1 and MN2 was 17.6 μm, D_Z between MN1 and MN2 was 0.5 μm, and D_{XYZ} between MN1 and MN2 (i.e., the radius of S2) was 17.607 μm.

[2] Accordingly, D_{XZY} between MN1 and MN3 was larger than D_{XYZ} between MN1 and MN2.

[3] At this focal depth D_{XY} between M1 and M2 was 15.708 μm.

[4] Accordingly, D_{XZY} between MN1 and MN4 was smaller than D_{XYZ} between MN1 and MN2.

[5] In the example shown in C, D_{XY} between MN1 and MN4 was 8.1 μm, D_Z between MN1 and MN4 was 7.8 μm, and D_{XYZ} between MN1 and MN4 (i.e., the radius of S4) was 11.245 μm.

[6] At this focal depth D_{XY} between M1 and M4 was 10.305 μm.

[7] In the example shown in D, D_{XY} between MN1 and MN6 was 8.3 μm, D_Z between MN1 and MN6 was 4.5 μm, and D_{XYZ} between MN1 and MN6 (i.e., the radius of S6) was 9.44 μm.

[8] At this focal depth D_{XY} between M1 and M6 was 5.317 μm.

[9] At this focal depth D_{XY} between M1 and M6 was 9.427 μm.

below the top surface of the section represented the upper and lower extension of S6. This guaranteed that the nearest neighbor of neuron #1 could indeed be found, and was not lying in a section adjacent to the investigated one. The neuron whose midpoint of the nucleolus came into focus at a depth of 25.44 µm below the top surface of the section could not be the nearest neighbor of neuron #1 because it was lying outside S6. Rather neuron #6 was the nearest neighbor of neuron #1 and the radius of S6 (i.e., 9.44 µm) was the nearest-neighbor distance of neuron #1. The Nearest-Neighbor Distance Distribution Function of the objects of interest within a given brain region can be obtained by combining the individual nearest-neighbor distances with the total number of objects of interest within a given brain region obtained with the Optical Fractionator and the volume of the region of interest obtained with the Cavalieri estimator.[1] Note that the MBF Bioscience Stereo Investigator software provides these analyses in an integrated manner and facilitates the determination of the nearest neighbor of a selected object of interest within a few seconds. Scale bar represents 25 µm.

To demonstrate the use of NNDDF analyses the spatial distribution of the neurons in the lateral nucleus of the amygdala in the right hemisphere from a human postmortem brain was determined (Figure 8). As already mentioned above, the volume of this nucleus was 384 mm^3 (estimated with the Cavalieri estimator), and the total number of neurons in this nucleus was 5.02×10^6 (estimated with the Optical Fractionator). The cumulated relative frequency distribution of the nearest-neighbor distances of n = 1,629 systematically and randomly sampled neurons in this nucleus is shown in Figure 9, part A, and the corresponding EDF plot in Figure 9, part B. From this figure, it could be concluded that the neurons in the lateral nucleus of the amygdala in the right hemisphere from the human postmortem brain shown in Figure 8 exhibited a clustered distribution.

The number of cells for which the nearest-neighbor distance should be determined in order to assess the spatial distribution of cells in a region of interest with NNDDF analysis is addressed in the section *How does consideration about variability affect the implementation of design-based stereological estimators?* on page 50 of this book.

There exists another approach to study the spatial distribution of cells in a region of interest: the Voronoi tessellation.[2] This approach focuses on the region of space that any cell within a region of interest occupies, i.e., the region of space that is closer to that cell than to any other. It has the shape of a polyhedron, called a Voronoi (or Dirichlet) polyhedron. In 2D, the Voronoi polyhedron is a polygon, the sides of which are located at mid-distance from the neighboring cells. The Voronoi polygons are contiguous and their set fills

---

[1] Schmitz et al., Cereb Cortex 2002;12:954.
[2] Duyckaerts et al., J Neurosci Methods 1994;51:47; Duyckaerts and Godefroy, J Chem Neuroanat 2000;20:83.

the space without intersection or overlap, i.e., they perform a *tessellation*. The use of Voronoi polygons provides information concerning spatial distribution as follows: the areas of the Voronoi polygons do not vary much when the cells are regularly distributed, whereas small and large polygons are found when cellular clusters are present. Unfortunately, it has not been possible so far to adapt algorithms that have been published for 3D tessellations to histological specimens.

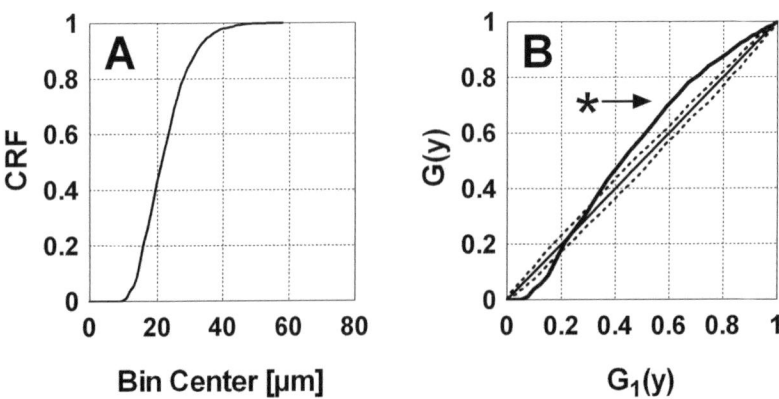

Figure 9. Results of the Nearest-Neighbor Distance Distribution Function (NNDDF) analysis of the spatial distribution of the neurons in the lateral nucleus of the amygdala in the right hemisphere from the human postmortem brain shown in Figure 8. A, cumulated relative frequency (CRF) distribution of the nearest-neighbor distances of n = 1,629 systematically and randomly sampled neurons in this nucleus. B, graphical comparison of this CRF distribution to corresponding distribution functions obtained from EDF plot analysis, modeling a virtual lateral nucleus of the amygdala in a human brain with the same volume and the same total number of neurons in the nucleus. Without going into detail, the EDF plot analysis showed that the neurons in the lateral nucleus of the amygdala in the right hemisphere from the human postmortem brain shown in Figure 8 exhibited a clustered distribution (asterisk in B). For a detailed description of the interpretation of NNDDF analyses with EDF plots see the on-line supplementary material for Schmitz et al. (Cereb Cortex 2002;12:954) at www.cercor.oupjournals.org/cgi/content/full/12/9/954/DC1.

What else can I do?

The most important design-based stereological methods that will be used in the majority of research analyses have already been covered. There are additional methods with specialized uses that we will not go into here. Among them are methods for estimating surface areas[1], and alternate methods for estimating length densities and total lengths of capillaries or cell processes[2], etc.

How can I assess the variation of estimates obtained with design-based stereology?

The results of quantitative histological investigations performed with design-based stereological methods are estimates of quantities rather than exact measurements. Since the results will vary if the same unbiased stereological procedure was to be independently repeated, it is important to understand the topic of variability of design-based stereological estimates. In practical applications of design-based stereology, the amount of sampling error (the difference between an estimate and the true value) is unknown. Therefore, several methods have been developed to predict the accuracy of a stereological estimate without the need to repeat the estimate multiple times. The results of these methods of prediction are usually presented in the form of a Coefficient of Error (CE).

The concept of the CE is distinct from that of biological variability (details in the next section) as well as from traditional statistical values relating to intrasubject variability found with repeated sampling (such as Variance and Standard Deviation). Usually CE calculations are intended to be a prediction of the Coefficient of Variation (CV) that would emerge if a region was indeed sampled many times. CV is a relative measure of dispersion independent of the magnitude of the mean.[3] Simply stated, with more sampling both the CV and the predicted CE will be reduced as more objects are counted.

Different CE formulas have been developed, based upon different assumptions and with different considerations taken into account, such as the shape of the region of interest, the distribution of objects within the region of interest, and the sampling criteria applied to the examination. Notably, the CE has no real biological meaning. Rather, it is most useful for evaluating the precision of stereological estimates.

---

[1] The corresponding method for isotropic uniform random sections has been developed by Sandau (*Spatial Grid*; Acta Stereol 1987;6:31), for vertical sections by Cruz-Orive and Howard (*Vertical Spatial Grid*; J Microsc 1995;178:146) and for (thick) sections with any plane of section by Kubinova and Janacek (*Isotropic Fakir Method*; J Microsc 1998;191:201).
[2] *L-cycloid Optical Fractionator for Lengths Estimates* (Stocks et al., J Neurocytol 1996;25:637).
[3] The CV is the ratio between the standard deviation of any values and the arithmetic mean of these values.

CE for volume estimates

Various methods have been developed to predict the CE for volume estimates obtained with the Cavalieri estimator.[1] In a pilot computer simulation, Schmitz and colleagues obtained CE values with the method given in Equation 4 in Roberts et al.[2], which were precise predictions of the actual coefficient of variation when investigating regular, "quasi-ellipsoidal" objects (C. Schmitz, personal communication). In addition, the method described in García-Fiñana et al.[3] is a refinement of Equation 4 of Roberts et al.[4], taking irregular objects into consideration. Therefore, this method is likely a good choice for predicting the CV of any estimates of volumes obtained with the Cavalieri estimator, provided the area estimates are based on point counting. It remains to be shown how this method can be used to compute CE values for volume estimates based on traced boundaries of the areas of a region of interest.

Calculating the CE of a volume estimate

In our example for estimating volume (beginning on page 21), the following numbers of intersections of the lattice and the cerebellar granule cell layer were counted: 1 intersection (point) on section no. 1, and 8 (no. 2), 33 (no. 3), 38 (no. 4), 51 (no. 5), 33 (no. 6), 38 (no. 7), 40 (no. 8), 40 (no. 9), 45 (no. 10), 32 (no. 11) and 11 (no. 12) intersections on the other sections, respectively. The CE of the corresponding estimate of the volume of the granule cell layer in the right cerebellar half of the investigated rat brain obtained with the method developed by García-Fiñana and colleagues[5] was 0.029.

If the CE is interpreted as a prediction of the CV of estimates that would be obtained if the same estimator were repeated *ad infinitum*, one can say as a rule of thumb that the estimated volume of the region

---

[1] Gundersen and Jensen, J Microsc 1987;147:229; Roberts et al., Br J Radiol 1994;67:1067; Geinisman et al., J Neurocytol 1996;25:805; García-Fiñana et al., Neuroimage 2003;18:505; Glaser, J Microsc 2005;218:1; Cruz-Orive and García-Fiñana, J Microsc 2005;218:6.
[2] Roberts et al., Br J Radiol 1994;67:1067.
[3] García-Fiñana et al., Neuroimage 2003;18:505.
[4] Roberts et al., Br J Radiol 1994;67:1067.
[5] García-Fiñana et al., Neuroimage 2003;18:505. Note that this method requires an estimate of the ratio between the average boundary length ($\overline{B}$) and the average area of the region of interest ($\overline{A}$) (i.e., $\overline{B}/\sqrt{\overline{A}}$). In this example the ratio $\overline{B}/\sqrt{\overline{A}}$ was estimated on n = 3 sections as 16.5.

of interest is with a probability of approximately 95% within ± 2 × CE of the true volume.[1]

In our example, the volume of the granule cell layer in the right cerebellar half of the investigated rat brain was estimated as 18.7 mm^3. Since the calculated CE was 0.029, the true value of the volume was with a probability of approximately 95% within the range 18.1 mm^3 to 19.3 mm^3.

CE for cell number estimates

For estimates of total numbers of cells obtained with the Optical Fractionator, various methods to predict the CE have been described.[2] Furthermore, three different methods have been proposed to predict the CE of corresponding estimates obtained with the $N_V \times V_{Ref}$ method[3]. Extensive computer simulations[4] have shown that you can precisely predict the CE of estimated total numbers of cells obtained with the Optical Fractionator depending on the spatial distribution of the cells.

---

[1] This is based on the idea that a) estimates from multiple unbiased samples of the same material would result in a Gaussian data set, b) roughly 95% of data points of a Gaussian data set fall within two standard deviations of the mean (true volume in this case) and c) this data set, when normalized by the mean (true volume) would graph the CV. The +/- 2*SD range would thus become a +/- 2*CV range (with CV now being expressed as a unitless percentage). If the CE is a good predictor of the CV, then there is an approximately 95% chance that the estimated volume is within +/- 2*CE of the true volume.

[2] Gundersen, J Microsc 1986;143:3-45; Gundersen and Jensen, J Microsc 1987;147:229; Cruz-Orive, J Microsc 1990;160:89; West et al., Anat Rec 1991;231:482; Thioulouse et al., J Microsc 1993;172:249; Scheaffer et al., Elementary Survey Sampling, 5th edition, PWS-Kent, Boston, 1996; Glaser and Wilson, J Microsc 1998;192:163; Larsen, J Neurosci Methods 1998;85:107; Schmitz, Anat Embryol 1998;198:371; Cruz-Orive, J Microsc 1999;193:182; Nyengaard, J Am Soc Nephrol 1999;10:1100; Glaser and Wilson, Acta Stereol 1999;18:15; Gundersen et al., J Microsc 1999;193:199; Schmitz and Hof, J Chem Neuroanat 2000;20:93.

[3] West and Gundersen, J Comp Neurol 1990;296:1; Geinisman et al., J Neurocytol 1996;25:805; Simic et al., J Comp Neurol 1997;379:482.

[4] Glaser and Wilson, J Microsc 1998;192:163; Schmitz, Anat Embryol 1998;198:371; Glaser and Wilson, Acta Stereol 1999;18:15; Schmitz and Hof, J Chem Neuroanat 2000;20:93.

Calculating the CE of a cell number estimate obtained with the Optical Fractionator

- If the spatial distribution of the cells resembles homogeneity or spatial randomness, you can apply the following methods: (i) "P6" in Figure 6 in Schmitz (Anat Embryol 1998;198:371), also given as Equation A5 in Glaser and Wilson (J Microsc 1998;192:163) and as Equation 24 in Nyengaard (J Am Soc Nephrol 1999;10:1100), (ii) the "Explicit Nugget Formula with m = 1" in Cruz-Orive (J Microsc 1999;193:182), or (iii) Equation 13 in Gundersen et al. (J Microsc 1999;193:199).

- In contrast, if the spatial distribution of the cells appears to be clustered, you can apply the following methods: (i) Chapter 4 in Scheaffer et al. (Elementary Survey Sampling, 5th edition, PWS-Kent, Boston, 1996), also given as Equation A4 in Larsen (J Neurosci Methods 1998;85:107) or described in the Appendix in Glaser and Wilson (J Microsc 1998;192:163), (ii) the "Implicit Nugget Formula with m = 1" in Cruz-Orive (J Microsc 1999;193:182), or (iii) Equation 15 in Gundersen et al. (J Microsc 1999;193:199).

An advantage of the methods described by Schmitz[1] and Schaeffer and colleagues[2] is that predictions of CE are substantially easier to calculate than using the other methods. If the spatial distribution of the cells under study resembles homogeneity or spatial randomness, a prediction of the CE of an estimated total number of cells with the Optical Fractionator is obtained by the reciprocal value of the square root of the number of counted cells. A disadvantage of the method described by Schmitz[3] is that it may under-predict the CV of estimated total numbers of cells if only a few sections are sampled.[4] The variability of cell counts depending on the number of sampled sections was recently investigated by Slomianka and West[5] as shown in Figure 10.

---

[1] Schmitz, Anat Embriol 1998;198:371.
[2] Schaeffer, et al. Elementary Survey Sampling, 5th edition, PWS-Kent, Boston, 1996.
[3] Schmitz, Anat Embriol 1998;198:37.
[4] Slomianka and West, Neuroscience 2005;136:757.
[5] Ibid.

Computer simulations have also shown that the methods mentioned above to predict the CE of estimated total numbers of cells obtained with the $N_V \times V_{Ref}$ method do not result in adequate predictions of CE.[1]

Methods to predict CEs for the remaining design-based stereological methods described in this book are currently not available.

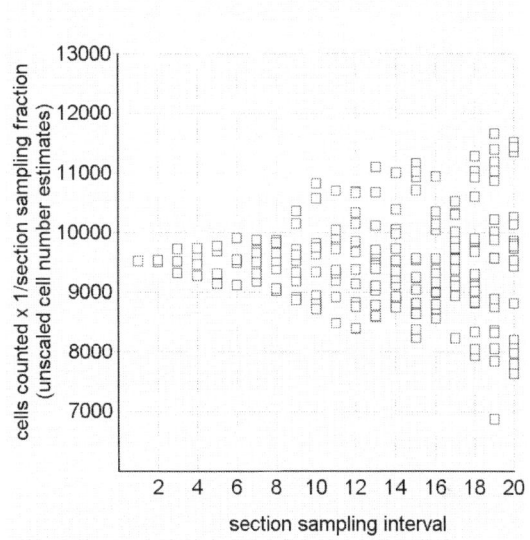

Figure 10. Variation of estimated total numbers of pyramidal cells in hippocampus area CA1 of the rat brain as a function of the section sampling interval.[2] Cells were initially counted using the Optical Fractionator on each section (i.e., the section sampling interval was 1). Then, the sections were subsampled by increasing the section sampling interval. For instance, a section sampling interval of 10 means that every tenth section was selected, yielding 10 different estimates of the total numbers of pyramidal cells (with the first estimate based on the analysis of the sections no. 1, 11, 21, ..., the second estimate based on the analysis of the sections no. 2, 12, 22, ..., and so on). As can be seen, the variation of the estimated total numbers of pyramidal cells increased when increasing the section sampling interval.

In summary, there is evidence from preliminary investigations that the variation of estimates of volumes obtained with the Cavalieri estimator can be precisely predicted. Furthermore, the variation of estimates of total numbers of cells obtained with the Optical Fractionator can be precisely predicted, provided the spatial distribution of the cells is (roughly) known. In any case, one of the most desirable advantages of design-based stereology over other

---

[1] Schmitz and Hof, J Chem Neuroanat 2000;20:93.
[2] Taken with permission from Slomianka and West, Neuroscience 2005;136:757.

quantitative histological techniques is that predictions about the variation of the estimated results are often possible.

How does consideration about variability affect the implementation of design-based stereological estimators?

A major domain of interest in design-based stereology has been the establishment of criteria for sampling requirements to achieve a specific level of variation in the estimates. Optimal stereological estimates are based on a sampling protocol that yields sufficient data to achieve the desired variation (i.e., precision) of the estimates, while making the work efficient by performing as few measurements as possible to achieve this precision.

Many researchers have decided to rely on a reasonable degree of oversampling (which is easy and fast when using automated stereology systems). For instance, numbers of counted cells have been increased to a range of 500 to 1,000 per individual.[1] In this manner, for homogeneously distributed cells, counting approximately 900 cells results in a CE of 0.033. The amount of time necessary to carry out these estimates is typically one day per region of interest, which is a reasonable compromise between the amount of time dedicated to the analysis and the precision of the obtained estimates. Likewise researchers have turned to analyzing volumes using the Cavalieri estimator on the investigation of at least 8 sections and at least 250 counted points (which typically requires approximately one hour per volume of interest), estimates of mean perikaryal or nuclear volumes on the investigation of a total of at least 500 cells (which typically requires no more than an hour when carried out simultaneously with estimates of total numbers of cells), and estimates of the average length of tubular objects on the investigation of at least 500 intersections of these objects with Space Balls (which typically requires about one day per region of interest).

Finally, oversampling is only one possible empirical solution to the problem of ensuring sufficient precision in an estimate (or group of estimates). For instance, the optimum sampling intensity in a study dealing with total numbers of cells depends not only on the spatial distribution of the cells within the tissue, but also on the kind of statistical analyses to be performed (i.e., comparison of group means, regression analysis, calculation of individual ratios of cell numbers, etc.). Because a general analytical solution to this problem does not exist, we recommend that someone knowledgeable in the field is involved in designing your individual sampling scheme before you start your

---

[1] See, for example, Schmitz et al., Mol Psychiat 2002;7:810; Bussière et al., Neuroscience 2003;117:577; Rutten et al., Neurobiol Aging 2007;28:91.

study. You should then verify your sampling scheme with a pilot study in which you perform oversampling so you can systematically examine the sampling parameters that will yield the most efficient and precise estimation in your design-based stereological study. Your final sampling scheme should address the number of animals or organs to be investigated (an issue not covered in this book), the number of sections to be analyzed, the sampling frequency on the sections to be examined, etc.[1]

---

[1] See also Table 2 in Schmitz and Hof, Neuroscience 2005;130:813.

PART IV

Application of Design-Based Stereology in Biological Research

In recent years, the application of design-based stereological methods to the analysis of the central nervous system as well as other organs has contributed considerably to our understanding of the functional and pathological morphology of the body. Among the many stereological studies in the literature, the recent ones described here are used as examples to demonstrate the broad range of current applications of this approach.

Application of design-based stereology in research on brain aging and Alzheimer's disease

Rapp and colleagues[1] investigated a well-characterized rat model of cognitive aging in combination with design-based stereological methods to test the hypothesis that age-related deficits in hippocampal learning are coupled with neuron death in the parahippocampal region. This hypothesis was based on the finding that damage involving the parahippocampal region causes significant learning and memory impairment in young subjects. However, the authors found that the total number of neurons in the entorhinal, perirhinal, and postrhinal cortices was largely preserved during normal aging. Furthermore, individual variability in hippocampal learning among the aged rats failed to correlate with neuron number in any region examined, and there was no indication of selective or disproportionate loss among the aged animals with the most pronounced cognitive impairment. Together with other data from the literature, these results indicated that age-related cognitive decline can occur in the absence of significant neuron death in any major, cytoarchitectonically defined component of the hippocampal system, suggesting that alterations in connectivity and other changes are more likely causative factors of age-related cognitive decline.

In addition, there is a considerable increase in postmenopausal years of life of women over the last 100 years, and age-related cognitive decline is among the most devastating aspects of brain aging. Synaptic proteins have been shown to be responsive to estrogen, and as such can provide useful markers of synaptic plasticity. The current understanding of estrogen's effects on synaptic plasticity in the brain is based primarily on data from rats, and it is well documented that estrogen increases dendritic spine density in hippocampal pyramidal cells of young—but not of aged—female rats. However, it is unknown whether the same is true for aging humans. The similarities in reproductive physiology and the pattern of female endocrine senescence between non-human primates and humans are substantial. Therefore, a non-human primate model offers an opportunity to employ a well-controlled, prospective, standardized ovariectomy and hormone replacement therapy that

---

[1] Rapp et al., Cereb Cortex 2002;12:1171.

mimics physiological cycles, isolates the effects of estrogen, and includes careful monitoring of estrogen levels, all of which have been difficult to achieve in human clinical studies. In this regard, Hao and colleagues[1] subjected young and old female rhesus monkeys to ovariectomy, followed by two administrations of either estradiol or vehicle with an interval of three weeks starting three months postovariectomy. Then, the total numbers of spinophilin-immunopositive dendritic spines within several compartments of the hippocampus were analyzed. The authors found that in both young and aged female monkeys, the estrogen treated groups had a significant increase in the number of spines by approximately 36% as compared to the untreated groups. These data have important implications for cognitive effects of estrogen replacement in postmenopausal women, and demonstrate that an estrogen replacement protocol that mimics normal physiological cycles with timed, intermittent peaks can have profound neurobiological effects. In a subsequent study, Tang and colleagues[2] showed on young monkeys an estradiol-induced 55% increase in the number of spines in layer I of prefrontal cortex area 46, whereas spine numbers in layer I of the opercular portion of primary visual cortex area V1 were equivalent across the two groups. These data suggest that estrogen's effects on synaptic organization influence select neocortical layers and regions in this non-human primate model, providing a morphological basis for enhanced prefrontal cortical functions following estrogen replacement.

Furthermore, Alzheimer's disease (AD) is characterized by a substantial degeneration of pyramidal neurons and the appearance of neuritic plaques and neurofibrillary tangles within specific regions of the brain. According to the "amyloid cascade" hypothesis for AD, amyloid is the cause of neurodegeneration in AD. To experimentally test this hypothesis *in vivo*, Schmitz and colleagues[3] studied a variety of transgenic mouse models[4] of AD with design-based stereological methods. Transgenic mice overexpressing the human β-cleaved C-terminal fragment of β-amyloid precursor protein (APP) with the I45F mutation under control of the prion protein promoter did not show age-related neuron loss in the hippocampus. By contrast, transgenic mice overexpressing both human mutant APP751 (carrying the Swedish and London mutations KM670/671NL and V717I, Thy1 promoter) and human mutant presenilin-1 (PS-1 M146L, HMG promoter) showed an age-related loss of

---

[1] Hao et al., J Comp Neurol 2003;465:540.
[2] Tang et al., Cereb Cortex 2004;14:215.
[3] Rutten et al., Neurobiol Dis 2003;12:110; Schmitz et al., Am J Pathol 2004;164:1495; Casas et al., Am J Pathol 2004;165:1289.
[4] A transgenic mouse is a mouse that carries a foreign gene that has been deliberately inserted into its genome. The foreign genes are usually constructed using recombinant DNA methodology, i.e., created artificially.

hippocampal pyramidal cells at 17 months of age. Neuron loss was observed at sites of extracellular Aβ aggregates (plaques) and surrounding astrocytes but, most importantly, was also clearly observed in brain areas distant from plaques. Finally, transgenic mice carrying knocked-in mutations in PS-1 (M233T and L235P) and overexpressing human mutant APP (KM670/671NL and V717I, Thy1 promoter) showed substantial neuron loss in hippocampus area CA1/2 by 50% on average already at 10 months of age. The average amount of extracellular Aβ aggregation was only 4.1% (volume / volume) in CA1/2, whereas in CA3 the average amount of extracellular Aβ aggregation was 6.8% and the average amount of neuron loss was only 21%. These data showed for the first time that in transgenic mouse models of AD, neuron loss occurs in relation to sites of extracellular Aβ aggregation but also independent of extracellular Aβ aggregation. Additional molecular and histological analyses confirmed the critical role of intraneuronal Aβ42 in neuron loss in these transgenic mouse models of AD. Overall, these findings have prompted important modifications of the original "amyloid cascade" hypothesis for AD.

Application of design-based stereology in schizophrenia research

In the human brain, the uncinate fasciculus interconnects the anterior temporal and inferior frontal lobes. Furthermore, the temporal lobes show a number of anatomical asymmetries, some of which are altered in schizophrenia. In this regard, Highley and colleagues[1] studied the area, fiber density, and total fiber number of left and right uncinate fasciculi in postmortem brains from schizophrenic patients and controls.[2] The authors found the uncinate fasciculus to be asymmetrical in both genders, being 27% larger and containing 33% more fibers in the right compared to the left hemisphere. There was no significant effect of schizophrenia on the uncinate fasciculus. The authors concluded that the uncinate fasciculus is larger in the right hemisphere, perhaps indicating greater right-sided frontotemporal connectivity. The unchanged size of the fasciculus in schizophrenia contrasted with commissural tracts, which were affected in the same brain series in a gender-specific manner.

In addition, the prefrontal cortex exhibits prominent functional, biochemical, and anatomical abnormalities in schizophrenic patients, but both postmortem studies and magnetic resonance imaging scans had not consistently revealed frontal volumetric deficits in schizophrenia. Accordingly, Selemon and colleagues[3] analyzed the total cortical gray and white matter

---

[1] Highley et al., Cereb Cortex 2002;12:1218.
[2] Using a modified counting frame, fiber intersections were counted.
[3] Selemon et al., Am J Psychiatry 2002;159:1983.

volumes, as well as frontal cortical gray and white matter volumes, of the left hemisphere in postmortem brains from schizophrenic patients and matched controls. The authors found a significantly smaller frontal gray matter volume in the schizophrenic patients compared to the controls (12% difference), whereas the differences between the groups in total gray and white matter volumes and frontal white matter volume (6%-8% smaller in the schizophrenic patients than in the controls) did not reach statistical significance. These data suggested that the neuropathology of the frontal lobe in schizophrenia may be more severe than that of the three other lobes, and may account for the prominence of prefrontal dysfunction associated with schizophrenia.

Application of design-based stereology in stem cell research

It has been known for decades that neurogenesis in the dentate gyrus of the hippocampus continues in the brain of adult rodents. To test the hypothesis that neurogenesis also occurs in the adult human hippocampus, Eriksson and colleagues[1] investigated the brains from patients who died from squamous cell carcinomas at the base of the tongue, in the larynx or in the pharynx, and had received bromodeoxyuridine (BrdU)[2] as intravenous infusion before death to assess the proliferative activity of the tumor cells.[3] Using a combined design-based stereology / confocal microscopy approach, the authors demonstrated that new neurons were generated from dividing progenitor cells in the dentate gyrus of adult humans. Specifically, approximately 22% of the BrdU-positive cells in the granule cell layer were also immunoreactive for NeuN, approximately 23% for neuron specific enolase, approximately 8% for calbindin, and approximately 18% for glial fibrillary acidic protein. These data demonstrated for the first time that neurogenesis occurs in the adult human brain, indicating that the human brain retains some potential for self-renewal throughout life.

Design-based stereological methods also played a pivotal role in studies showing that mice living in an enriched environment[4] had more granule cells in the dentate gyrus of the hippocampus[5], and that this phenomenon was mediated by a survival-promoting effect that is selective for neurons rather than increased proliferation of progenitor cells.[6]

---

[1] Eriksson et al., Nat Med 1998;4:1313.
[2] BrdU is a thymidine analog, is incorporated into the DNA of dividing cells (i.e., it labels DNA during the S phase of the cell cycle) and can be detected immunohistochemically in their progeny.
[3] The proliferative activity of the tumor cells was expressed as BrdU labeling index.
[4] Enriched environment means to switch from standard housing to conditions with opportunities for social interaction, exploration, and physical activity.
[5] Kempermann et al., Nature 1997;386:493.
[6] Kempermann et al., J Neurosci 1998;18:3206.

Design-based stereological methods were also applied in subsequent studies focusing on the molecular mechanisms regulating adult neurogenesis in the dentate gyrus. For instance, O'Kusky and colleagues[1] investigated the total number of neurons in the dentate gyrus of transgenic mice overexpressing insulin-like growth factor-I (IGF-I) postnatally in the brain to test the hypothesis that IGF-I is involved in the growth and development of the hippocampal dentate gyrus. Compared to nontransgenic littermate controls, the authors found significantly more neurons in the dentate gyrus of the transgenic mice (55% more neurons on postnatal day 14, and 61% on postnatal day 130). These results indicated that IGF-I promotes neurogenesis in the developing hippocampus in vivo. Another example is the discovery of protein tyrosine kinases (PTKs) and protein tyrosine phosphatases (PTPs) as important regulators of neurogenesis in the adult hippocampus in recent years. Specifically, PTKs mediate the stimulating action of growth factors on neurogenesis and are counterbalanced by PTPs. In this regard, Bernabeu and colleagues[2] hypothesized that downward regulation of the leukocyte common antigen-related (LAR) PTP function in progenitor cells of the hippocampus might provide a novel strategy to promote neurogenesis. The authors tested this hypothesis by injecting 3-month-old LAR-/- mice[3] with BrdU and determined the number of BrdU positive cells as well as the total number of neurons in the hippocampus at different time points after 6 days of BrdU administration. At 1 day and 4 weeks following the last BrdU administration, LAR-/- mice showed a significant increase in cells in the hippocampus immunoreactive for both BrdU and NeuN, consistent with increased neurogenesis. Furthermore, LAR-/- mice displayed a significant 37% increase in the total number of neurons in the hippocampus granule cell layer compared to normal littermate controls. This study introduced LAR as the first PTP found to be expressed in progenitors of dentate gyrus granule cells, and pointed to inhibition of LAR as a potential therapeutic strategy to promote neurogenesis.

Application of design-based stereology in research on the lung, kidney and placenta

There is a large body of literature dealing with the use of design-based stereology in the investigation of organs other than the brain. In this section, we look at some highlights.

---

[1] O'Kusky et al., J Neurosci 2000;20:8435.
[2] Bernabeu et al., Mol Cell Neurosci 2006;31:723.
[3] In LAR-/- mice, a β-galactosidase/neomycin transgene was inserted into the LAR gene, resulting in efficient disruption of the gene without obvious effects on normal embryonic development (Skarnes et al., Proc Natl Acad Sci USA 1995;92:6592).

With respect to the lung, chronic hypoxia caused by migration of native coast dwellers to high altitude or by chronic lung disease can lead to the development of increased pulmonary vascular resistance, followed by pulmonary hypertension. It is obvious that altitude-induced hypertension does not offer benefit and may indeed be maladaptive. It was a long-standing hypothesis that hypoxia-induced loss of small blood vessels in the lung plays an important role in the pathogenesis of pulmonary hypertension. This hypothesis, however, was in contrast to the well-known potent angiogenic effect of chronic hypoxia in all other vascular beds including the one in the brain.[1] To overcome this situation, Howell and colleagues[2] subjected rats to either normal oxygen conditions (inspired O_2 fraction = 0.21) or hypoxia (inspired O_2 fraction = 0.10) for two weeks. Afterwards they investigated the lungs of the animals with a variety of design-based stereological methods combined with confocal microscopy. The authors found that chronic hypoxia resulted in increased total pulmonary vessel length, volume, and endothelial surface area, as well as in an increased number of endothelial cells. This was the first report demonstrating hypoxia-induced angiogenesis in the mature pulmonary circulation. The authors concluded that the widely accepted paradigm of hypoxia-induced loss of small vessels as a key structural change contributing to the development of pulmonary hypertension in high altitude adaptation and chronic lung disease must be revised. In contrast, it appeared that hypoxia-induced angiogenesis in the mature pulmonary circulation may have important beneficial consequences for gas exchange in the lung.

Another seminal design-based stereological study in the framework of kidney research focused on the pathogenesis of primary ("essential") hypertension. Supported by a substantial body of experimental data it had been hypothesized that having relatively fewer nephrons[3] in the kidneys renders an otherwise healthy person more susceptible to renal disease, hypertension, or both. Corresponding data in humans, however, have been largely circumstantial and indirect. This situation changed when Keller and colleagues[4] compared the total number and volume of glomeruli in middle-aged patients with a history of primary hypertension or left ventricular hypertrophy (or both) and renal arteriolar lesions with the number and volume of glomeruli in normotensive subjects matched for age, gender, height, and weight. The authors found that the patients with hypertension had significantly fewer glomeruli per kidney than the matched normotensive controls. Furthermore, the patients with hypertension had a significantly greater glomerular volume

---

[1] For review see LaManna et al., J Exp Biol 2004;207:3163.
[2] Howell et al., J Physiol 547:133.
[3] Nephrons are the basic structural and functional units of the kidney, composed of an initial filtering component (the glomerulum) and a tubule specialized for reabsorption and secretion (the renal tubule).
[4] Keller et al., N Engl J Med 2003;348:101.

than the controls. These data were the first direct evidence in humans supporting the hypothesis that the number of nephrons is reduced in patients with primary hypertension.[1]

Finally, restricted fetal growth can significantly increase the risk of mortality, neurodevelopmental handicap, and other morbidities in the neonatal stage and in childhood. Furthermore, small size at birth in human population studies has been found to be associated with increased risk of high blood pressure and abnormal glucose tolerance in adulthood.[2] These associations can be reproduced in animal models of restricted growth *in utero*. Thus, a better understanding of the control of fetal growth would make a significant impact on the burden of common diseases. In this regard it has been hypothesized that restricted fetal growth is directly associated with alterations in the capacity of the placenta to supply nutrients, and that specific imprinted genes[3] can regulate nutrient supply by the placenta. Sibley and colleagues[4] specifically tested the hypothesis that the gene coding for the insulin-like growth factor 2 (IGF-2)[5] transcribed from the placental-specific promoter regulates the development of the diffusional permeability properties of the mouse placenta. To this end, the authors investigated mice in which placental-specific IGF-2 was deleted (P0). At embryonic day 19 (E19), placental and fetal weights in P0 mice were reduced to 66% and 76%, respectively, compared to wild type mice. Design-based stereological analysis of histological sections revealed that compared to wild-type mice, the surface area of the exchange barrier in the labyrinth of the placenta was reduced in P0 mice, while its thickness was increased. As a result, the average theoretical diffusing capacity in P0 knockout placentas was dramatically reduced to 40% of that of wild-type placentas. By measuring the transfer of three inert hydrophilic solutes of increasing size (^{14}C-mannitol, ^{51}CrEDTA, and ^{14}C-inulin) in vivo, the authors found that in the P0 mutants, the permeability surface area product for the tracers was reduced to approximately 70% of that of wild-type controls at E19, independent of the size of the tracers. This study demonstrated that placental IGF-2 regulates the development of the mouse placenta's diffusional exchange characteristics, thereby providing a mechanism for the role of imprinted genes in controlling

---

[1] For a review specifically focusing on design-based stereological investigations of the kidney see Nyengaard, J Am Soc Nephrol 1999;10:1100.
[2] See, for example, Barker et al., N Engl J Med 2005;353:1802.
[3] Imprinted genes are genes whose expression is determined by the parent that contributed them. Accordingly, imprinted genes violate the usual rule of inheritance that both alleles in a heterozygote are equally expressed. Imprinted genes play a crucial role in placental development and fetal growth.
[4] Sibley et al., Proc Natl Acad Sci USA 2004;101:8204.
[5] The insulin-like growth factor II gene is paternally expressed in the fetus and placenta (Constancia et al., Nature 2002;417:945).

placental nutrient supply and fetal growth. Accordingly, altered placental IGF-2 could be a cause of idiopathic intrauterine growth restriction in humans. [1]

---

[1] For a review specifically focusing on design-based stereological investigations of the placenta see Mayhew, Placenta 2006;27 Suppl:17.

PART V

The Future of Design-Based Stereology

In recent years, we have seen that design-based stereology has become the standard investigative method in quantitative histology. While these inquiries provided new and valuable information, they have also helped to identify current limitations in design-based stereology. In this section, we will look at some projects that are collaborations between researchers and MBF Bioscience, which look for solutions to these limitations.

Development of novel design-based stereological methods

Novel areas of study may require novel methods. The quantitative analysis of the cerebral cortex cytoarchitectural features is one area. The following are two examples.

First, there is evidence for the existence of a general principle of anatomical and functional microstructure of the cerebral cortex, with the latter based on its modular organization.[1] In Nissl-stained tissue the visibility of these modules, usually called cell minicolumns, depends on the linear arrangement of pyramidal cells and the existence of cell-free space on both sides of the column core (Figure 11).[2] It has been proposed that a cell minicolumn is a complex processing and distributing unit that links a number of inputs to a number of outputs via overlapping internal processing chains.[3] Importantly, alterations in these cellular microdomains in the cerebral cortex may be crucial in understanding the neurobiological deficit in neuropsychiatric diseases.[4] For instance, in Nissl-stained tissue from patients with autism it was found that cell minicolumns in the cortex are narrower than in controls, with less peripheral neuropil space and increased spacing between their constituent cells.[5] However, the current concept of cell minicolumns is based on computerized analysis of two-dimensional (2D) images from brain sections rather than using the three-dimensional (3D) investigations that have become possible with the advent of design-based stereology (as explained in detail in Part II and Part III).

---

[1] Hutsler and Galuske, Trends Neurosci 2003;26:429.
[2] Buxhoeveden and Casanova, Brain 2002;125:935.
[3] Mountcastle, Brain 1997;120:701.
[4] Casanova et al., Ann Neurol 2002;52:108; Casanova et al., Psychiatry Res 2005;133:1.
[5] Casanova et al., Neuroscientist 2003;9:496.

Figure 11. Modular organization of the cytoarchitecture of the cerebral cortex (cellular microdomains or cell minicolumns), illustrated for prefrontal area 9 of the cerebral cortex from the same human postmortem brain described in Figure 7 and Figure 8 (500 μm thick, gallocyanin-stained coronal section obtained from tissue embedded in gelatin, deeply frozen at -60°C and then serially sectioned using a cryomicrotome). The image shows all cortical layers from the pial surface on top of the image to the border between gray and white matter on the bottom of the image. Note the clearly visible, vertical arrangement of cells and the existence of cell-free space on both sides of the column core. The picture was generated from 240 images captured with a 40x lens (Olympus Plan Apo; NA = 1.00), made into one montage using the MBF Bioscience Virtual Slice module of the Stereo Investigator software. Scale bar represents 200 μm.

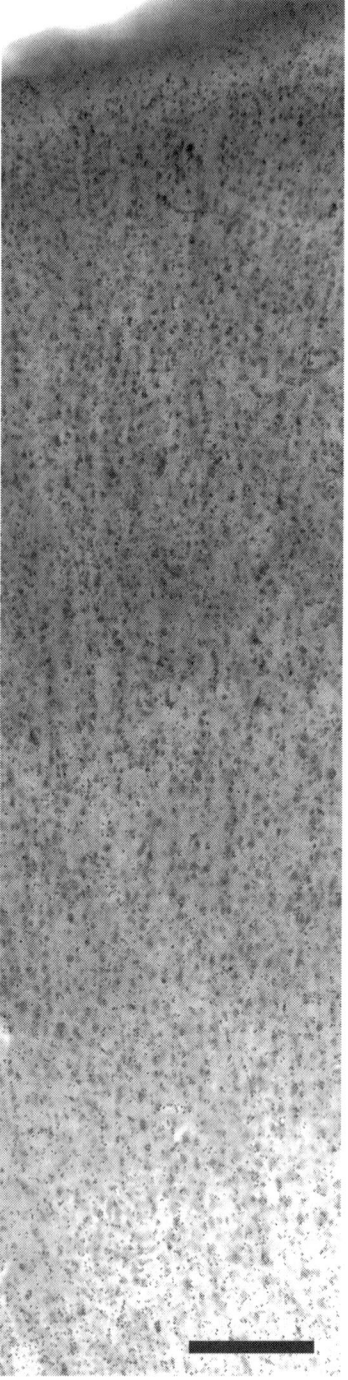

Accordingly, several authors have regarded claims about cell minicolumns with skepticism.[1] Notably, the questions raised do not argue against the usefulness or importance of the cellular microdomain concept, but rather argue for further analysis of these anatomical features and their possible functional implications. Research is underway to address cell minicolumns with a novel approach, facilitating the analysis of linear arrangements of cells in brain sections in 3D. In doing so, the analysis of cellular microdomains will become independent of the plane of section, the staining intensity of the tissue and the shape of the cells, all being potential sources of bias in analyzing cellular microdomains with optical density measurements by computerized imaging in 2D.

Second, hippocampal pyramidal cell disarray was found in postmortem brains from patients with schizophrenia and autism,[2] possibly pointing to disturbed neuronal migration during brain development.[3] However, the corresponding analyses were carried out in 2D, introducing potential bias due to the plane of section of the investigated tissue. Novel approaches to study pyramidal cell disarray in 3D will be launched soon.

More precise prediction of the variability of stereological estimates

New developments are underway to precisely predict the variability of estimates of (i) volumes obtained with the Cavalieri estimator, (ii) mean volumes of biological objects obtained with the Nucleator, the Rotator or with Point Sampled Intercepts, (iii) the total length of tubular objects of interest obtained with the Space Balls method and (iv) the spatial distribution of biological objects with the Nearest-Neighbor Distance Distribution Function method. Another improvement in this regard will be the development of new strategies to design stereological sampling protocols so that both sufficient precision of the estimates is achieved and the work is efficient by performing as few measurements as necessary to achieve this precision.

Integration of design-based stereology with anatomical mapping

The advanced integration of anatomical mapping with design-based stereological analyses is another promising area of development. As outlined in

---

[1] For details see Jones, Proc Natl Acad Sci USA 2000;97:5019; Buxhoeveden and Casanova, Brain 2002;125:935; Rockland and Ichinohe, Exp Brain Res 2004;158:265.
[2] Conrad et al., Arch Gen Psychiatry 1991;48:413; Mukaetova-Ladinska et al., Neuropathol Appl Neurobiol 2004;30:615.
[3] Jonsson, Eur Arch Psychiatry Clin Neurosci 1997;247:120.

Part III, each stereological analysis starts with the identification of the boundaries of the region of interest. In many cases this is very easy, as for example shown in Figure 1 for the different, clearly identifiable layers of the rat cerebellum on Nissl-stained sections. You just trace the boundaries of the regions of interest on the images displayed on a computer monitor.[1] In many situations, however, another approach is necessary. For instance, one of the most reliable methods available for identifying boundaries of medial and dorsolateral prefrontal, orbitofrontal, and retrosplenial cortical areas in the human brain is based on the identification of neurons immunoreactive for a certain non-phosphorylated neurofilament protein (SMI-32).[2] If you wish to apply this method to delineate the region of interest and perform stereological analyses on adjacent, differently stained sections, it is necessary to exactly copy the boundaries identified on the sections immunoprocessed for the detection of SMI-32 to the adjacent sections. Another example is the identification of cortical regions in the human brain on thick, Nissl-stained sections based on cytoarchitectural criteria, followed by stereological analysis of the capillary length in these regions on adjacent sections immunoprocessed for the detection of collagen IV.[3] Perhaps the most important development in this area will be the availability of software interfaces integrating the identification of the boundaries of brain regions of interest with the Gray Level Index (GLI) method[4] and stereological analyses on the same sections within the identified regions.

Integration of design-based stereology with confocal microscopy

Integrating design-based stereology with confocal microscopy is another exciting area of development. One of the major—and frustrating—constraints on the rigorous use of design-based stereology in quantitative histology is that the various protocols of cutting-edge research require double- and triple immunofluorescence labeling and microscopic image stacks with high contrast and high axial resolution, which can only be achieved with confocal microscopy.[5] These research areas include the fate of newborn and transplanted cells,[6] the accumulation of intraneuronal Aβ in vivo in animal models of Alzheimer's disease (AD),[7] the composition of proteins in

---

[1] Glaser and Glaser, J Chem Neuroanat 2000;20:115.
[2] Hof et al., J Comp Neurol 1995;359:48; Nimchinsky et al., J Comp Neurol 1997;384:597; Vogt et al., J Comp Neurol 2001;438:353.
[3] Kreczmanski et al., Acta Neuropathol 2005;109:510.
[4] Schleicher and Zilles, J Microsc 1990;157:367; Schleicher et al., J Chem Neuroanat 2000;20:31; Schleicher et al., Anat Embryol 2005;210:373.
[5] Peterson, Methods 1999;18:493.
[6] Rakic, Nat Rev Neurosci 2002;3:65.
[7] Casas et al., Am J Pathol 2004;165:1289.

presynaptic boutons in the brain or in neuromuscular junctions,[1] and the cell-type-specific analysis of alterations in gene expression in the vicinity of and at distance from pathological hallmarks such as extracellular Aβ aggregates in AD[2] or experimentally induced lesions in the brain.[3] Accordingly, relatively few confocal microscopic design-based stereological analyses have been performed so far (for instance Lazarov et al., 2006).[4] In general, the necessary confocal image stacks could be manually generated with laser scanning confocal microscopes, if equipped with the appropriate hardware and software necessary to achieve systematic random sampling. However, without this equipment, the tremendous effort required by the observer, together with photobleaching and potential errors associated with the manual generation of thousands of confocal image stacks necessary in a given study, made it virtually impossible to perform confocal design-based stereological analyses without a semi-automated, computer-based confocal microscopic stereology system.[5]

This unsatisfactory situation changed with the recent launch by MBF Bioscience of the Stereo Investigator Confocal Spinning Disk (SI-SD) system (Figure 12), as well as tightly integrated support between Stereo Investigator and the Olympus Fluoview laser scanning confocal systems.[6] The heart of the SI-SD system is a customized spinning disk unit (DSU; Olympus, Tokyo, Japan) which generates the confocal illumination. This is achieved by a spinning disk with a pattern of slits that creates a virtual pinhole as the disk spins at 3,000 revolutions per minute. It features different disk configurations designed to optimize the trade-off between confocality and light throughput for a wide range of objectives, and it uses arc lamp illumination for maximum excitation wavelength flexibility at reasonable cost. The system is fully motorized, allowing computer software to automatically move the disk in and out of the lightpath (for confocal vs. non-confocal viewing) and to select fluorescent filter sets via the included computer-controlled emission and excitation filter wheels. The DSU optical design allows excellent performance from near UV to near IR. Confocal images can be collected with objectives from 2× to 100×. The DSU is attached to a motorized microscope to facilitate automated Z-stack and 3-D image acquisition. Furthermore, the system can be equipped with a computer-controlled lens turret in addition to the three-axis high accuracy computer-controlled stepping motor specimen stage, linear Z-axis position encoder, ultra-

---

[1] Rutten et al., Am J Pathol 2005;167:161.
[2] Burbach et al., Glia 2004;48:76.
[3] Lazarov et al., J Neurosci 2006;26.
[4] Ibid.
[5] In this regard it is important to note that the advent of semiautomated, computer-based stereology systems in widefield microscopy was the prerequisite for design-based stereology becoming a practical laboratory method.
[6] The MBF Bioscience SI-SD and laser confocal systems are currently the first and only systems facilitating confocal microscopic design-based stereological analyses in quantitative histology.

high sensitivity cameras, and software for confocal design-based stereological analyses. This set-up not only allows maximum acquisition speed, but also minimal photobleaching of fluorophores, which is of paramount importance in acquisition of multi-channel digital image stacks from a high number of microscopic fields in the same tissue section (Figure 13).

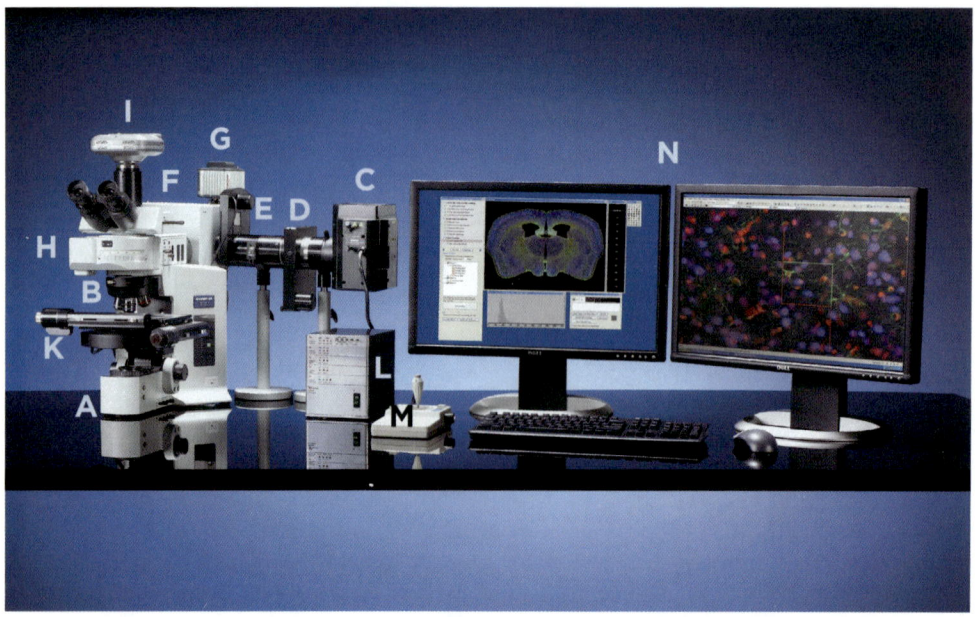

Figure 12. The MBF Bioscience Stereo Investigator Confocal Spinning Disk (SI-SD) system. **A**, Olympus BX51 microscope. **B**, Computer controlled lens turret. **C**, White light source (mercury lamp). **D**, Computer-controlled filter changer (excitation side). **E**, Computer-controlled shutter. **F**, Spinning disk unit with confocal disk and computer-controlled filter changer (emission side). **G**, Ultra-high sensitivity monochrome back-thinned electron multiplier CCD camera. **H**, Computer-controlled filter turret. **I**, Color digital CCD camera with switchable monochrome mode. **K**, Three-axis highest accuracy computer-controlled stepping motor specimen stage. **L**, Control unit. **M**, Three-axis joystick. **N**, PC workstation with high-end video card, controlling software and 2 × 24" WUXGA LCD monitor. The linear Z-axis position encoder is not shown.

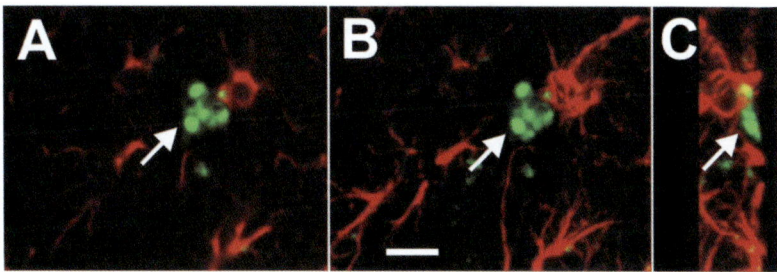

Figure 13. Representative example of the image obtained with the MBF Bioscience Stereo Investigator Confocal Spinning Disk (SI-SD) system during a confocal microscopic design-based stereological investigation. The panels show immunohistochemical detection of glial fibrillary acidic protein (GFAP; red) and bromodeoxyuridine (BrdU; green)[1] within the dentate gyrus of a 24-month-old mouse (30 μm thick coronal cryostat section)[2] Photomicrographs were taken from the same microscopic field at 24 consecutive focal planes below the upper surface, with a distance of 1 μm between the focal planes. **A,** Two-dimensional representation of the microscopic field in XY view at a distance of 8 μm below the upper surface of the section. **B** and **C,** Three-dimensional reconstruction (maximum intensity projection) of the image stack in XY view (B) and YZ view (C). A BrdU immunoreactive cell was found that showed no double-immunoreactivity for GFAP (arrow). No adjustments of contrast and brightness were made, and no deconvolution was performed. Note that the image channels (red and green in this example) have to be analyzed consecutively when the stereological analyses are performed on-site on the near real-time images displayed on the video screen. Alternatively, the image channels can be analyzed simultaneously when stereological analyses are performed off-site on the generated digital image stacks. Scale bar represents 10 μm.

If single or multi-channel confocal image stacks are acquired using this system, design-based stereological analyses can be performed either on-site (at the confocal microscope), or off-site based on the generated data files and digital image stacks (freeing up the SI-SD system for other investigators). This latter feature of allowing semi-automated image stack collection over multiple slides, sections, and regions of interest, followed by offline analysis, makes this system ideal for both high throughput and also for shared core equipment facilities.

The SI-SD system can also perform confocal design-based stereological analyses using real-time imaging on a computer monitor. This is the first time it

---

[1] The mouse received five daily injections with BrdU (50mg/kg body weight) for five consecutive days six weeks before killing.
[2] Free-floating sections. Primary antibodies: monoclonal mouse anti-BrdU 1:1000 (Sigma, St. Louis, MO, USA) and polyclonal rabbit anti-GFAP 1:1600 (DAKO, Glostrup, Denmark). Secondary antibodies: donkey anti-mouse IgG Alexa Fluor 488 (1:100; Molecular Probes, Eugene, OR, USA) and donkey anti-rabbit IgG Alexa fluor 594 (1:100; Molecular Probes).

has been possible to do real-time confocal stereology, i.e., confocal stereology, without first acquiring image stacks.[1]

For visual inspection of tissue sections prior to confocal design-based stereological analysis, which includes defining the region of interest, evaluating any regional differences in fluorescent signal, archiving overview images in brightfield or fluorescence illumination modes, etc., the SI-SD system can optionally include a computer-controlled 'front' filter turret that can accept up to 5 different filter cubes, and a color digital camera with switchable monochrome mode. This camera allows real-time inspection of tissue in color, or in monochrome into the near IR. With these optional settings, the SI-SD system can be used as a normal bright-field or fluorescent microscope when not using the DSU illumination mode.

If full integration of a stereology system is required with a laser scanning confocal microscope, MBF Bioscience also has a number of options for a fully integrated package in conjunction with an appropriately equipped Olympus Fluoview 300, 500, or Fluoview 1000 confocal microscope system.

Integration of design-based stereology with digital representations of histological tissue specimens

With most modern design-based stereology systems, histological tissue specimens are viewed through a microscope using a camera, and the analyses are performed in real-time on-site via images displayed on a computer monitor. While the results of these investigations are valid and reliable, the integration of design-based stereology with confocal microscopy (outlined above) has highlighted some of the limitations to this approach:

- Immunofluorescently labeled sections can suffer from fading of the fluorescent dyes, which causes difficulties in long-term archiving of the corresponding sections.

- High-end confocal microscopes are expensive, which limits the availability of confocal microscopic design-based stereology systems in research laboratories.

- The NIH have repeatedly emphasized the need for improved collaboration in research by making original image data available to external researchers after publication.

---

[1] In confocal mode, this is limited to a single fluorophore at a time.

Three-dimensional (3D) Virtual Tissue slides will enable researchers to overcome these limitations. A 3D Virtual Tissue slide is defined here as a data file containing a Z-series of single focal-plane image montages. Each focal plane represents an image montage of the full XY extent of a tissue section (or, at the user's discretion, a selected part of the tissue) where each tile of the montage is a photograph of a single field of view taken at a user-selected magnification. A number of such focal planes are acquired and examined together to allow the user to dynamically follow structures such as cells, cell processes and vessels through their full 3D extent within the Virtual Tissue slide. The ultimate goal of Virtual Tissue is to replicate the navigational feel and possibilities of a microscope, while adding analysis and annotation tools only possible with a digital slide file.

Two-dimensional (2D) virtual slide acquisition systems, slide storage databases, slide servers, and internet slide viewers from MBF Bioscience are already in use in a number of academic and telemedicine institutions. These systems have overcome the traditional obstacles of accurate image montage acquisition (with proper background correction, image stitching, etc.), the huge size of these data sets, their storage, and real-time viewing at any selected magnification. 2D Virtual Tissue with selected parts of the tissue represented in 3D ready for design-based stereological analyses is also already available (Figure 14), as well as 3D Virtual Tissue as a stand-alone product.

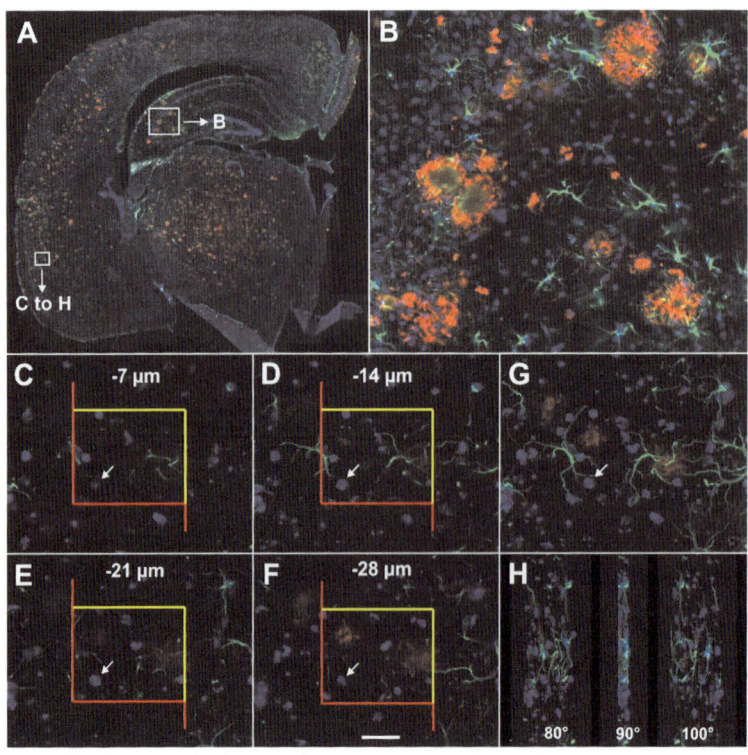

Figure 14. Virtual Tissue generated from a 30 μm thick coronal cryostat section of the brain from a transgenic mouse model of Alzheimer's disease.[1] The panels show immunohistochemical detection of glial fibrillary acidic protein (GFAP; green) and beta-amyloid (red) as well as counterstaining with Hoechst (blue).[2] **A**, Two-dimensional representation of a Virtual Tissue slide made of 4,602 (78×59) blue-, green- and red-channel images of the same microscopic fields in XY, and 60 consecutive focal planes with a distance of 0.5 μm between the focal planes. The tissue was imaged using a 60× objective; images were acquired with spinning disk confocal illumination mode. **B**, High-power image taken at the position of the upper small rectangle indicated in A. **C** to **F**, High-power images showing the same microscopic field (at the position of the lower small rectangle indicated in A) at four consecutive focal planes, with a distance of 7 μm between the focal planes. An unbiased counting frame is shown at each focal plane, representing an unbiased virtual counting space

---

[1] 17-month old transgenic mouse overexpressing both human mutant amyloid precursor protein (APP) 751 (carrying the Swedish and London mutations KM670/671NL and V717I, Thy1 promoter) and human mutant presenilin-1 (PS-1 M146L, HMG promoter).

[2] Free-floating sections. Primary antibodies: monoclonal mouse anti-GFAP (1:1600; Sigma, St. Louis, MO, USA) and rabbit anti-mouse polyclonal antiserum 730 against human Aß and P3 (1:5000; gift from Dr. Gerd Multhaup, Berlin, Germany). Secondary antibodies: donkey anti-mouse IgG Alexa Fluor 488 (1:100; Molecular Probes, Eugene, OR, USA) and donkey anti-rabbit IgG Alexa fluor 594 (1:100; Molecular Probes).

with a height of 21 µm. The arrow indicates a cell within the unbiased virtual counting space that comes into focus at the focal plane represented in C, and is also found in D to F. **G** and **H**, Three-dimensional (3D) reconstructions (maximum intensity projections) of the same 3D Virtual Tissue in XY view (G) and YZ view (H; shown in three different angles as indicated). The arrow in G points to the same cell indicated in C to F. Only minor adjustments of contrast and brightness were made, which in no case altered the appearance of the original materials. Importantly, no deconvolution was performed. Scale bar represents 25 mm in A, 1.6 mm in B, and 100 µm in C to H. Note that when the sample contains green material, the counting frame is shown in red and yellow.

The full integration of design-based stereology with 3D Virtual Tissue will result in the following benefits and advantages to the field of design-based stereology:

- Each analysis, whether evaluations of cell numbers, cell size, lengths of capillaries etc., is associated with an image file. Researchers can, for the first time, share their quantitative data obtained with design-based stereology as well as the full 3D image context (i.e., functional virtual equivalents of the original tissue slides) from which it was obtained. This data can be either shared remotely (in which case the researcher has control of all access to the data set) or it can be sent to others on digital media.

- Virtual Tissue can be acquired using the highest magnification objective lens on a microscope allowing for access to both high and low-resolution views of the tissue section from the same source file (Figure 14). You can use any suitably equipped microscope[1] to generate Virtual Tissue image sets. High-resolution views allow design-based stereological analyses, while medium, low-resolution and 'macro' views of anatomical organization (derived from the high-resolution image tiles) allow for observations that would normally be available using lower power objectives, or which would not be available on a research microscope at all. This real-time multi-resolution viewing is achieved by a) storing Virtual Tissue files in a special multi-resolution pyramidal file format, and b) only displaying the tiles for the necessary field of view from the required resolution file level, c) allowing both fast navigation and high image quality without burdening networks or individual computers with needs for high capacity or bandwidth.

- Performing analyses on 3D Virtual Tissue slides allows for a more efficient use of resources as the microscope is freed from quantification

---

[1] A modern upright biological research microscope with an MBF Bioscience Virtual Tissue acquisition system installed.

functions. Furthermore, researchers without access to full-scale stereology systems can still perform design-based stereological analyses in their own laboratories.

- It can be difficult to verify findings from physical slides. If researchers want to verify earlier findings—either from their own research or from another laboratory—they must gain access to the physical slides used in the original study. With a permanent digital record in the form of a 3D Virtual Tissue slide and the ability to share Virtual Tissue remotely in real-time via a data network, slide material can be examined and analyzed at any time and from any remote location.[1] This also enables researchers to revisit their own material with fresh insights gained from new findings, and to perform additional analyses at will on the same tissue, anytime and anywhere. For the user, this can often also eliminate the effort required to reproduce experimental tissue for additional analysis. Furthermore, this feature allows immediate feedback from collaborators and consultants.

- Virtual Tissue slides, or access to them, can be provided for electronic publication with journal articles, vastly increasing the amount of information presented with each study. Researchers can have original slide material instantly available to anyone in any remote location.

- Finally, 3D Virtual Tissue slide technology supports the development of new automated object detection analyses to speed research throughput and to increase the scope of studies. For instance, efficient, image-wide exact quantification on vast amount of data (e.g., 100,000 × 100,000 pixels or more per optical plane) is time consuming when performed manually; it therefore may only become feasible with automated algorithms. Since an acquired (montaged) image can be used for object recognition and other automated analysis functions, 3D Virtual Tissue permits entire tissue sections to be analyzed in a single step. This will enhance studies focusing on image-wide quantifications such as the spatial distribution of cell attributes, cell clusters, and colocalization analyses, along with reproducible extraction of more data for higher statistical validity.

---

[1] The remote location would require standalone stereology software in order to perform additional stereological analyses based on the Virtual Tissue slide.

PART VI

Further Reading

Books dedicated to design-based stereology:

Bertram J, Nurcombe V, Wreford N. (1999) *Stereological Methods for Biology.* London: International Thomson Publishing Services.

Elias H, Hyde DM. (1983) *Guide to Practical Stereology.* Basel: Karger.

Howard V, Reed MG. (2005) *Unbiased Stereology: Three-Dimensional Measurement in Microscopy.* New York: Taylor and Francis.

Mouton PR. (2002) *Principles and Practices of Unbiased Stereology: An Introduction for Bioscientists.* Baltimore, MD: The Johns Hopkins University Press.

Reith A, Mayhew TM. (1988) *Stereology and Morphometry In Electron Microscopy: Some Problems & Their Solutions.* London: CRC Press.

Russ JC, Dehoff RT. (2000) *Practical Stereology* (2nd ed). New York: Kluver Academic / Plenum Publishers.

Vedel Jensen EB. (1998) *Local Stereology.* London: World Scientific Publishing Company.

Weibel ER. (1979) *Stereological Methods, Vol. 1. Practical Methods for Biological Morphometry.* London: Academic Press.

Weibel, ER. (1980) *Stereological Methods, Vol. 2. Theoretical Foundations.* London: Academic Press.

Reviews dedicated to design-based stereology:

Calhoun ME, Mouton PR. (2001) Length measurement: new developments in neurostereology and 3D imagery. *J Chem Neuroanat 21,* 61-69.

Coggeshall RE. (1999) Assaying structural changes after nerve damage, an essay on quantitative morphology. *Pain Suppl 6,* S21-S25.

Coggeshall RE, Lekan HA. (1996) Methods for determining numbers of cells and synapses: a case for more uniform standards of review. *J Comp Neurol 64,* 6-15.

Cruz-Orive LM, Weibel ER. (1990) Recent stereological methods for cell biology: a brief survey. *Am J Physiol 258,* L148-L156.

Dorph-Petersen KA, Nyengaard JR, Gundersen HJ. (2001) Tissue shrinkage and unbiased stereological estimation of particle number and size. *J Microsc 204,* 232-246.

Geuna S. (2000) Appreciating the difference between design-based and model-based sampling strategies in quantitative morphology of the nervous system. *J Comp Neurol 427,* 333-339.

Glaser J, Glaser EM. (2000) Stereology, morphometry, and mapping: the whole is greater than the sum of its parts. *J Chem Neuroanat 20,* 115-126.

Gundersen HJ. (1986) Stereology of arbitrary particles. *J Microsc 143,* :3-45.

Gundersen HJ. (1992) Stereology: the fast lane between neuroanatomy and brain function–or still only a tightrope? *Acta Neurol Scand Suppl137,* 8-13.

Gundersen HJG, Bagger P, Bendtsen TF, Evans SM, Korbo L, Marcussen N, *et al.* (1988) The new stereological tools: Disector, fractionator, nucleator and point sampled intercepts and their use in pathological research and diagnosis. *Acta Pathol Microbiol Immunol Scand 96,* 857-881.

Gundersen HJ, Bendtsen TF, Korbo L, Marcussen N, Moller A, Nielsen K, *et al.* (1988) Some new, simple and efficient stereological methods and their use in pathological research and diagnosis. *Acta Pathol Microbiol Immunol Scand 96,* 379-394.

Howell K, Hopkins N, Mcloughlin P. (2002) Combined confocal microscopy and stereology: a highly efficient and unbiased approach to quantitative structural measurement in tissues. *Exp Physiol 87,* 747-756.

Hyman BT, Gomez-Isla T, Irizarry MC. Stereology: a practical primer for neuropathology. (1998) *J Neuropathol Exp Neurol 57,* 305-310.

Kreczmanski P, Schmidt-Kastner R, Heinsen H, Steinbusch HWM, Hof PR, Schmitz C. (2005) Stereological studies of capillary length density in the frontal cortex of schizophrenics. *Acta Neuropathologica 109,* 510-518.

Kubinova L, Janacek J, Karen P, Radochova B, Difato F, Krekule I. (2004) Confocal stereology and image analysis: methods for estimating geometrical characteristics of cells and tissues from three-dimensional confocal images. *Physiol Res 53 Suppl 1,* S47-S55.

Mayhew TM. (1992) A review of recent advances in stereology for quantifying neural structure. *J Neurocytol 21,* 313-328.

Mayhew TM. (1996) How to count synapses unbiasedly and efficiently at the ultrastructural level: proposal for a standard sampling and counting protocol. *J Neurocytol 25,* 793-804.

Mayhew TM. (2006) Stereology and the placenta: where's the point? – a review. *Placenta 27 Suppl A,* S17-S25.

Mayhew TM, Gundersen HJ. (1996) 'If you assume, you can make an ass out of u and me': a decade of the disector for stereological counting of particles in 3D space. *J Anat 188,* 1-15.

Oorschot DE, Peterson DA, Jones DG. (1991) Neurite growth from, and neuronal survival within, cultured explants of the nervous system: a critical review of morphometric and stereological methods, and suggestions for the future. *Prog Neurobiol 37,* 525-546.

Peterson DA. (1999) Quantitative histology using confocal microscopy: implementation of unbiased stereology procedures. *Methods 18,* 493-507.

Roberts N, Puddephat MJ, McNulty V. (2000) The benefit of stereology for quantitative radiology. *Br J Radiol 73,* 679-697.

Royet JP. (1991) Stereology: a method for analyzing images. *Prog Neurobiol 37,* 433-474.

Schmitz C, Hof PR. (2000) Recommendations for straightforward and rigorous methods of counting neurons based on a computer simulation approach. *J Chem Neuroanat 20,* 93-114.

Schmitz C, Grolms N, Hof PR, Boehringer R, Glaser J, and Korr H. (2002) Altered Spatial Arrangeent of Layer V Pyramidal Cells in the Mouse Brain following Prenatal Lowdose X-Irradiation. A Stereological Study using a Novel Three-dimensional Analysis Method to Estimate the Nearest Neighbor Distance Distribution of Cells in Thick Sections. *Cereb Cortex 12*, 954-960.

Schmitz C, Hof PR. (2005) Design-based stereology in neuroscience. *Neuroscience 130*, 813-831.

Tandrup T. (2004) Unbiased estimates of number and size of rat dorsal root ganglion cells in studies of structure and cell survival. *J Neurocytol 33*, 173-192.

Von Bartheld C. (2002) Counting particles in tissue sections: choices of methods and importance of calibration to minimize biases. *Histol Histopathol 17*, 639-648.

West MJ. (1993) New stereological methods for counting neurons. *Neurobiol Aging 14*, 275-285.

West MJ. (2001) Stereological methods for estimating the total number of neurons and synapses: issues of precision and bias. *Trends Neurosci 22*, 51-61.

West MJ. (2001) Design based stereological methods for estimating the total number of objects in histological material. *Folia Morphol 60*, 11-19.

West MJ. (2002) Design-based stereological methods for counting neurons. *Prog Brain Res 135*, 43-51.

Part VII

Glossary

Anisotropic: Having properties that differ according to the direction of measurement. Compare with **isotropic**.

Area sampling fraction: The ratio of the counting frame's area to the area formed by the fractionator sampling grid. Often abbreviated as *asf*.

B: Symbol used to represent the base area of the counting frame.

Bias: A statistical sampling or measurement error caused by systematically favoring some outcomes over others. Bias causes the mean of the estimated values to deviate from the true value.

Cavalieri estimator: An unbiased stereological method that estimates the volume of a structure from individual parallel cross-sectional areas (typically sections) employing Cavalieri's principle. Typically, the areas are estimated using a point-counting method.

Cell density: The number of cells per unit volume.

Coefficient of Error: Abbreviated as CE. The precision of a population size estimate. Calculated as the standard deviation of the sample divided by the mean of the sample.

Coefficient of Variation: Abbreviated as CV. Calculated as the standard deviation of the population divided by the population mean.

Counting frame: A two-dimensional stereological probe that is used with specific counting rules to count or select particles. The counting frame is a rectangle with extensions of two infinite rays. It is typically displayed using the colors red and green, which assist in implementing the counting rules. Use of a counting frame along with the counting rules results in all particles having an equal probability of being selected, regardless of shape, size, orientation, and distribution.

Counting space: The 3D volume contained within the Optical Disector probe. It is calculated by the counting frame area multiplied by the height of the Optical Disector.

Cut section thickness: The section thickness as measured directly from the sectioning device (cryostat, microtome, etc.). This is the thickness of a section prior to histological processing, which may cause shrinkage in the z axis. Also known as the block advance of the microtome.

Cycloids: A sine weighted curve that is used in some stereological probes in vertically sectioned material.

Cytoarchitecture: The arrangement of cells in the brain, especially the cerebral cortex.

Design-based stereology: An unbiased stereological technique using sampling design. This method eliminates the need for making assumptions about the size, shape, or orientation of objects.

Disector: A stereological probe for counting or selecting objects using a pair of adjacent physical sections. The method uses a counting frame with counting rules to determine the objects that are counted. The Disector is also referred to as the Physical Disector, especially in literature that follows the introduction of the Optical Disector.

Empirical distribution function plot: A graphical representation of the proportion (or frequency) of values less than or equal to each value.

Exclusion lines: One of a set of lines that comprise a counting frame. The counting frame is created from 2 sets of lines at right angles to create a rectangle. If an object crosses the set of lines designated as exclusion lines, the object is not counted. Exclusion lines are typically displayed as red or dashed lines. Compare with **inclusion lines**.

Fractionator: A systematic random sampling method that selects a portion of a region of interest. The fractionator principle is used in many areas of design-based stereology. Fractionator based sampling schemes are unbiased.

Geometric probe: A geometric shape, typically a set of lines, points, cycloids, or curves, which is overlaid on an object to investigate and obtain quantitative information of the object.

Gray Level Index method: A methodology for preprocessing an image to automatically identify boundaries between cytoarchitectonically defined brain regions.

H: Symbol denoting the height of the counting space or Optical Fractionator.

Inclusion lines: One of a set of lines that comprise a counting frame. The counting frame is created from 2 sets of lines at right angles to create a rectangle. If an object crosses the set of lines designated as inclusion lines, or lies within the counting frame the object is counted. Inclusion lines are typically displayed as green or solid lines. Compare with **exclusion lines**.

Isotropic: The property of being identical in all directions. Compare with **anisoptopic**.

Isotropic uniform random (IUR) sections: Sections cut in a manner that they fulfill the criteria to be both isotropic (having no preferred orientation) and a uniformly distance (interval) apart. The orientation of the sections is selected at random. Note, the sectioning methods used by most researchers, i.e., sagittal, coronal, etc., have preferred orientations and do not result in IUR sections.

Köhler illumination: A method of microscopy that provides optimum specimen visualization using brightfield illumination by focusing the light from the microscope condenser at the level of the specimen.

Lost caps, plucked cells: Artifact caused by sectioning methods where cells near the top surface (lost caps) or bottom surface (plucked cells) of a section are removed by the blade of the microtome.

Merz: A stereological probe designed to estimate the length of objects.

Model based stereology: Stereological methods, which assume that objects have particular size, shape, and orientation which can be approximated using a mathematical model (such as the method of Abercrombie (1946)). Model based methods are biased unless the objects exactly match the model. Until the advent of design-based stereology, model based stereology was the only form of stereology.

Mounted section thickness: The thickness of tissue sections after histological processing.

Nearest-Neighbor Distance Distribution Function (NNDDF): A relative frequency distribution of the location of cells in a given volume.

Nucleator: A design-based estimator that uses the intersection of rays and the cell surface for the estimation of the volume and cross-sectional area of cells or small objects (nuclei, plaques, etc.).

Number-weighted mean volumes: (arithmetic mean volume): Generally known as the mean volume, unambiguously stating that the mean is weighted by the number of particles rather than the volume of the particles (volume weighted mean). Compare with **Volume-weighted mean volumes**.

$N_v \times V_{Ref}$: Stereological method used to calculate the total number of cells within a region of interest. Uses the fractionator method for assessing mean cell density, which is then multiplied by the total volume of the region.

OF: See **Optical Fractionator**.

Optical Disector: A stereological probe for counting or selecting objects in a tissue section. This is an extension to the basic Disector method, which is applied to a thick section using a series, or stack, of Disectors. Rather than using pairs of physical sections (the basic Disector method), optical sectioning is used with a thin focal plane throughout the depth of the section for determining if objects fall within the Disector.

Optical Fractionator: A design-based stereological method using a two-stage systematic sampling method that is used to estimate the number of objects in a specified region of an organ. This method combines the Optical Disector method with the fractionator sampling method. Like the Physical Fractionator, the Optical Fractionator is typically used when the population is too large to count exhaustively. Compare with **Physical Fractionator**.

Physical Disector: A stereological probe for counting or selecting objects using a pair of adjacent physical sections. The method uses a counting frame with counting rules to determine the objects that are counted. The Physical Disector is also often referred to as the Disector, especially in literature that precedes the introduction of the Optical Disector.

Physical Fractionator: A design-based stereological method using a two-stage systematic sampling method that is used to estimate the number of objects in a specified region of an organ. This method combines the Physical Disector method with the fractionator sampling method. Like the Optical Fractionator, the Physical Fractionator is typically used when the population is too large to count exhaustively. Since the Physical Fractionator relies on the Physical Disector method, pairs of thin sections are required. Compare with **Optical Fractionator**.

Point Sampled Intercepts: A method for efficiently estimating the volume-weighted mean volume of cells or small objects.

Probe: A geometric shape, typically a set of lines, points, cycloids, or curves, which is overlaid on an object to investigate and obtain quantitative information of the object.

Profile counting: The act of counting the cross-sections of an object rather than the entire object. When an object, such as a cell is cut, the profile of the cross-section is visible. Objects counted using a profile counting method in a single 2D "representative section" are susceptible to sampling bias.

Randomized start: A random initial section is chosen within the sampling interval to begin the section sampling using a random number generator. The use of a randomized start is critical to systematic random sampling.

Representative section: Acting as a true example of an entire region. In reality, upon close inspection sections that initially appear to be "representative sections" often are not actually representative of the entire region.

Rotator: A method for estimating the mean number-weighted particle volume, similar to the Nucleator. Also referred to as the Planar Rotator.

Section sampling fraction: The known interval of sections sampled through an object of interest (e.g., examining 1 in every 5 sections has a section sampling fraction of 1/5). Often abbreviated as *ssf*.

Shrinkage: The difference between the original section thickness (section cut thickness) and that after it has been histologically prepared for stereology (mounted section thickness).

Space Balls: A design-based stereological probe that estimates length of tubular objects such as dendrites or blood vessels in thick sections. The method works by counting the number of transections of the objects to be counted with a virtual sphere. Also known as Isotropic Virtual Spheres.

Stereology: The method of quantifying 2D and 3D structures using estimation methods.

Systematic random sampling (SRS): Regular sampling (using a known interval) that begins with a randomized start. SRS is the basis for design-based stereological sampling methods, such as the Fractionator.

t: Abbreviation for the average mounted thickness of the section.

Thickness sampling fraction: The proportion of the section height that is sampled. Calculated by dividing the mounted section thickness by the height of the counting space. It is also known as the height sampling fraction. Often abbreviated as *tsf*.

True population: The actual total population of the region of interest.

UCF, unbiased counting frame: Also known as the counting frame. Compare with **Counting frame**.

Uniform distance between sections: The known section sampling interval, generally a multiple of the section cut thickness.

Vertical Sectioning: A method of producing sections from a specimen that contains a natural flat surface, i.e., bone. A vertical axis is selected that is perpendicular to the surface. The material is randomly rotated about the vertical axis. Sections are then cut parallel to the vertical axis.

Virtual counting space: The volume associated with the counting space or box. See **Optical Disector**.

Volume-weighted mean volumes: The mean object volume if the objects are weighted (sampled) proportional to their volume. Compare with **Number-weighted mean volumes**.

Voronoi tessellation: Partitioning of plane using a set of points where in each cell partition, all points in the cell are closer to the point that defines the cell than to any other point.

INDEX

aging, 55
Alzheimer's disease, 56
anatomical mapping, 17, 67
anisotropy, 16
area sampling fraction, 32
asf. *See* area sampling fraction
B, 32
bias, 4, 23, 30, 31, 39, 67, 82
Cavalieri, 16, 21, 23, 24, 37, 40, 43, 46, 49, 67
Cavalieri estimator, 15, 22, 39, 43, 50
cell density, 3, 7, 8, 24
cell fragments, 7, 23
cell profiles. *See* cell fragments
coefficient of error, 24, 45, 46, 47, 48, 49, 50
coefficient of variation, 45, 46, 47, 48
confocal microscopy, 58, 60, 68, 72, 80, 81
confocal spinning disk, 69
counting frame, 6, 18, 24, 25, 26, 28, 30, 41, 57, 74
counting space, 24, 28, 30, 75
cresyl violet, 3, 9
cycloids, 36
design-based stereology, iii, iv, vii, 3, 4, 7, 8, 9, 15, 16, 17, 26, 31, 45, 49, 50, 55, 57, 58, 59, 65, 67, 68, 69, 72, 75, 79, 80
Dirichlet, 43
isotropic, 16, 33, 34, 36, 39, 45
isotropic uniform random sections, 16, 33, 36
isotropic virtual planes, 36
isotropy, 16
kidney, 59, 60, 61
Köhler illumination, 23
lost caps, 31
lung, 59, 60
Merz, 36

Nearest-Neighbor Distance Distribution Function, 15, 16, 17, 40, 41, 44, 67
neurogenesis, 58, 59
Nucleator, 17, 33, 34, 35, 67
$N_V \times V_{Ref}$, 8, 24, 47, 49
Optical Disector, 25, 27, 28, 29, 30
Optical Fractionator, 8, 15, 17, 18, 24, 31, 33, 36, 37, 40, 43, 45, 47, 48, 49
Optical Rotator, 33
Physical Fractionator, 17, 24
placenta, 59, 61, 62, 81
Planar Rotator, 35
plucked cells, 31
Point Sampled Intercepts, 17, 33, 34, 36, 67
Purkinje cells, 3, 4, 6, 7, 8, 9, 10, 21, 26, 28, 30, 32, 35, 36
rat cerebellum, 3, 4, 7, 9, 27, 29, 68
representative section, 3
Rotator, 17, 33, 34, 35, 67
schizophrenia, 57, 67
section sampling fraction, 32
Space Balls, 15, 17, 36, 37, 38, 39, 50, 67
ssf. *See* section sampling fraction
stem cell, 58
stereology, 1, 3, 63, 79, 80, 81
systematic random sampling, 17, 24, 33, 69
t, 3, 10, 32
thickness sampling fraction, 32
tsf. *See* thickness sampling fraction
UCF. *See* unbiased counting frame
unbiased counting frame, 6, 18, 24, 25, 26, 28, 30, 35, 41, 74
unbiased estimation, 8
unbiased virtual counting space, 7, 8, 9, 15, 24, 25, 27, 28, 29, 30, 31, 32, 74

vertical sections, 16
virtual sphere, 37

Virtual Tissue, 73, 74, 75, 76
Voronoi tessellation, 43

About the Authors

Jack Glaser is the President and co-founder of MBF Bioscience where has led the development of MBF's products, including Neurolucida and Stereo Investigator, MBF's stereology system. He works closely with many leading international researchers to advance the fields of computer microscopy and quantitative analysis.

Geoff Greene is Chief Scientific Applications Officer for MBF Bioscience and is engaged in product development, customer support, and sales. He was one of the original authors of Neurolucida and Stereo Investigator. He holds a B.S. in Mathematics from the University of Vermont.

Susan J. Hendricks, Ph.D. is a Staff Scientist at MBF Bioscience. She has a broad interest in developmental neuroscience with a focus in neuroanatomy. Her dissertation research at the University of Virginia focused on the development of the rat gustatory system in the laboratory of David L. Hill. Postdoctoral training with Edwin W Rubel at the University of Washington and Rae Nishi at the University of Vermont expanded her expertise in imaging and quantitative methods in the developing auditory system of the chick. Joining MBF Bioscience in 2005, she assumed the role of staff scientist, assisting product development and educational training in stereology.